Rudolf Steiner and the Atom

Rudolf Steiner

and the

Atom

by
Keith Francis

Adonis Press

copyright © 2012 by Keith Francis

Published by Adonis Press
321 Rodman Road
Hillsdale, New York 12529
www.adonispress.org

ISBN 978-0-932776-44-0

Cover design and lettering by Dale Hushbeck

All rights reserved
Printed in the United States of America

Author's Note

This book is intended for people who have some knowledge of the work of Rudolf Steiner (1861-1925), the Austrian philosopher, scientist, educator, and seer who founded, among other endeavors, the Anthroposophical Society and the Waldorf School Movement; but I have tried to make it accessible to those who are new to anthroposophy. And if you have no specialist knowledge of physics or mathematics, don't be scared when you see an equation – there are only a few and you can just read around them!

Contents

Author's Note..v
Introduction..ix

I. The Atom - A Historical Background..........................1
    (i)   Prelude in Greece..1
    (ii)   Elements and Principles...................................9
    (iii) The Way of Truth..12
    (iv) Atoms...17
    (v)   Roadblock...20
    (vi) Atoms back in Vogue......................................25
    (vii) Making Waves..34
    (viii) Rudolf Steiner meets the Atom.......................35
    (ix) Rejection...38
    (x)   The Age of Electricity.....................................40
    (xi) The Electrical Atom and Human Thought........44

II. A Background for Quanta..50
    (i)   Origins..50
    (ii)   Thermal Radiation..53
    (iii) Enter Max Planck...65

III. Steiner in the Quantum Age....................................80
    (i)   Physical Science and Spiritual Science...........80
    (ii)   The Goethean Alternative..............................86
    (iii) The Primal Phenomenon..............................98

IV. Bohr's Atom – Antecedents...................................106
    (i)   Periodic Tables..106
    (ii)   From Siberia with Love.................................113
    (iii) Predictions and Confusions..........................117
    (iv) The Hydrogen Spectrum..............................122

(v) Cathode Rays..124
(vi) The Unstable Atom......................................130

V. The Rutherford-Bohr Atom..........................135
(i) Bohr Gets Involved..135
(ii) The Hydrogen Atom......................................138
(iii) Beyond Hydrogen...140

VI. Late Words from Rudolf Steiner.....................150
(i) A Science of Dead Matter.............................150
(ii) The Demonic Atom..153
(iii) Don't be an Ostrich!.....................................158
(iv) The Struggle for Human Consciousness..........160
(v) So what about the Electron?.........................169

VII. The Atom After Steiner.................................176
(i) Waves and Particles......................................176
(ii) Knabenphysik...181
(iii) "Thou Shalt Make No Mental Image."............182
(iv) Discontinuities and Probabilities..................186
(v) HBJ or the Three-Man-Paper........................190
(vi) Schrodinger's Wave Mechanics.....................193
(vii) Indeterminacy..196
(viii) Quantum Physics and the Periodic Table........201
(ix) More about Probability................................202
(x) Niels Bohr – A Goethean Physicist?.............207
(xi) Are Particles Real?......................................212

VIII. Epilogue...215

Appendix..246
Endnotes..251
Bibliography..265
About the Author..268

Introduction

(i)

In the year 2012, talking about "the atomic theory" is like talking about "the theory that the earth is round." Something that everyone knows to be true is not referred to as a theory. Everything is made of atoms, which contain electrons and a nucleus made of protons and neutrons, which are made of quarks, which are probably made of something else... Apart from this widely accepted piece of lore, most people hardly give a thought to atoms from one day's end to the next, but the information that Rudolf Steiner didn't believe in the atomic theory is apt to bring the response, "Well, if he were alive now, he'd have to believe in it, wouldn't he?" One of the purposes of this book is to examine this proposition. After all, the atomic theory that Steiner rejected throughout his life is very different from the one that most people accept now.

Or is it? This question arises not because there is any doubt that the current views of professional physicists differ radically from those familiar to Steiner when he gave his most mature assessments of atomic physics, but for the far simpler reason that most people have very little idea of what physicists actually think about atoms today.

Steiner received his scientific education at a time when the atom was a unitary, indestructible particle and, at the time of his death, it still consisted of only two kinds of particle, the heavy, positive proton and the light, negative electron; but it is arguable that the modern profusion of particles does not in itself make atomic physics a radically different thing from what it was a hundred years ago. In 1911, the year in which Rutherford discovered the nucleus, physicists could still apply ordinary electrical and mechanical laws to atomic particles and hope to get the right answers. What most people who know that everything is made of atomic particles don't know is that, by the time of Steiner's death in 1925, a revolution was in progress that had already shown that this hope was unfounded. Atomic physics required a different kind of mathematics and a different kind of mechanics, and involved the recognition that the old ideas of particles and waves had to be radically transformed. The electron and the proton, which had been unknown in Steiner's youth, lost the kind of predictable, objective existence that they had seemed to possess in his middle age.

These changes had only a marginal effect on popular ideas of atomic theory. Throughout the twentieth century, a person who was decently educated, but not a professional physicist or a diligent amateur, was still likely to think of the atom as it was conceived in the period from 1913 to 1920, with a central nucleus and shells of orbiting electrons. This concept remained in use as a teaching tool and popular image, and the fact that it had to a large extent lost its credibility as a picture of reality was not generally known. The atomic physics that may or may not have been acceptable to Steiner if he had lived longer is neither the kind that existed during his lifetime nor the kind that still exists in many people's heads today; so to ask whether he would have changed his mind is not quite on the point. It was the physicists who changed their minds and we should ask, instead, "Would Steiner have responded more positively to the new theory than he did to the old one?"

The question is of great interest but it is also hypothetical, so although I shall eventually make my opinion clear, I have a second purpose in mind. What I hope to do is to give students of Rudolf Steiner's work a background in the history of the atom and Steiner's thoughts about it that will enable them to form a meaningful relationship to modern scientific theory. Where did the idea of atoms come from in the first place, what kind of a thing was the atom when Steiner encountered it as a schoolboy in 1872, and how did it develop over the course of his life and in the following years? How was it that throughout his life he viewed atomism as a "rash hypothesis" and atomic concepts as the results of mental inertia, while still speaking of the profound esoteric and moral implications of the atom as though it actually existed in the physical world? What did he see as the influence of atomistic thinking on the future of the human race, and how does the situation look now, almost ninety years after his last remarks about the atom?

(ii)

Modern physics has its vulnerabilities, of which the great physicist Max Born gives some indication.

> *In 1921 I believed – and I shared this belief with most of my contemporary physicists – that science produced an objective knowledge of the world, which is governed by deterministic laws. The scientific method seemed to me superior to other, more subjective ways of forming a picture of the world – philosophy, poetry and religion; and I even thought the unambiguous language of science to be a step towards a better understanding between human beings.*

In 1951 I believed none of these things. The border between subject and object had been blurred, deterministic laws had been replaced by statistical ones, and although physicists understood one another well enough across all national frontiers, they had contributed nothing to better understanding of nations, but had helped in inventing and applying the most horrible weapons of destruction.

I now regard my former belief in the superiority of science over other forms of human thought and behavior as a self-deception due to youthful enthusiasm over the clarity of scientific thinking as compared with the vagueness of metaphysical systems.[1*]

This transformation in Born's thinking gives a compelling picture of the revolution in atomic physics that took place between the last few years of Steiner's life and the aftermath of the Second World War. Since that time there have been many striking new developments, but there has been no fundamental change of outlook comparable to the one that took place in the nineteen-twenties and thirties and provides much of the subject matter of the second half of this book.

Born still believed in the usefulness of science in the search for truth and a better life and was perfectly clear that the change in its view of phenomena was "not arbitrary but forced on the physicists by their observations. The final criterion of truth is the agreement of a theory with experience..."

We shall hear from Born from time to time and realize that, like many of his colleagues, he was a person of great worth and moral fiber; and we shall have to ask the questions, "What kind of theory?", "What kind of experience?", and "Truth about what?"

---
* See Endnotes, beginning on page 253.

(iii)

I shall mention the quantum several times before we reach the denouement of the story of its invention, so it may be helpful to give a preliminary indication of what kind of a thing a quantum is – or, perhaps, was. Elementary texts tell us that a quantum is the smallest quantity of something that it is possible to have. We can't have less than one atom of hydrogen and, according to Max Planck, we can't have less than one quantum of radiant energy. The concept of minima is not new – twenty-three hundred years before Planck, Aristotle showed that there are grave philosophical difficulties in the idea that substance is infinitely divisible. In the early twentieth century, Planck and Einstein showed that the phenomena of thermal radiation and the photoelectric effect placed apparently insurmountable obstacles in the way of treating radiant energy as a continuous stream that is infinitely divisible. Later developments in quantum theory are responsible for my personal definition, which is that the quantum is the smallest quantity that a physicist can imagine.

(iv)

One of the most remarkable things about a coastline is that the more closely you examine it, the longer it gets. What looks, on a small-scale map, to be a smooth curve, appears on a larger scale as a jagged, pock-marked line, so that if you actually walked along the shore you might cover ten times the distance you expected. You can shorten the distance by flying over all the promontories, coves, inlets and estuaries but only at the expense of missing the most interesting scenery. A similar dilemma is inherent in the history of atomic science and, perhaps, of everything else. You may feel that there are great over-arching principles with such broad and far-reaching implications that much of the detail becomes superfluous; or you may be dogged, diligent and foolhardy enough to follow all the ins and outs, risking the loss not only of any kind of perspective but also of your reader's attention. I don't believe anyone has yet come up with the kind of insight that would make knowledge of the actual twisted, knotted, convoluted course of history unnecessary and unhelpful. Steiner certainly provided an essential outline, but his indispensable accounts of vital stages in the evolution of human consciousness and the cosmic struggles for control of it strongly suggest complexity as the general rule.

In order to understand and appreciate the physical science of the twentieth century, you must stand beside the physicists and chemists who created that science, and fight their battles with them, at least in your imagination. To make this possible, while doing my best to avoid over-stressing the reader, I have followed a course somewhere between the extremes of over-simplification and over-elaboration.

The earliest known references to atoms appeared in the writings of certain Greek philosophers in the fifth century B.C. Philosophy, as we know it, had only recently come into existence and my book opens with a consideration of the esoteric reasons for its appearance at that time and the philosophical problems that seemed to make some kind of atomic theory necessary.

Chapter I
The Atom – A Historical Background

(i)
Prelude in Greece

One thing that most people seem to agree about is that human nature has always been the same: "You can't change human nature." The ancients wore funny clothes and were ignorant of human biology, calculus and other useful disciplines, but their thoughts, perceptions, feelings and desires were really just the same as ours. Rudolf Steiner found this assumption to be deeply mistaken and, out of his research into the esoteric background of human history, gave a very different picture, in which the descent of the human race is the story of a long and very gradual separation from its spiritual origin. Events of world history are outward signs of an evolving human consciousness and of struggles in the spiritual world for the control of that evolution. The following sketch of the origins and development of atomic theories begins at a time when the evolution toward an independent human consciousness had already come a long way.

*

The history of the atom goes back at least 2,500 years. It seems to have made its initial appearance in fifth century BC Greece and to have percolated into Indian and Chinese culture shortly thereafter. The tale of the early days of the concept of the atom and the controversies that swirled around it among the early Greek philosophers is an important episode in the development of Greek philosophy, which is in turn part of the much larger story of the evolution of consciousness.

The earliest Greek philosophers appeared in the sixth century BC, which was roughly six hundred years after the siege of Troy and two or three hundred years after the composition of the Iliad. Their emergence was also about a century after the lifetime of a grumpy Boeotian farmer called Hesiod, whose father had left him and his brother a small plot of land in a hamlet called Ascra at the foot of Mount Helicon, the traditional home of the Muses. Hesiod referred to Ascra as "a cursed place, cruel in winter, hard in summer, never pleasant," and spent his long winter evenings writing the poetry from which later Greek writers derived a great deal of their mythology. His *Theogony* concerns the origins of the world and of the gods, and shows a special interest in their family relations. According to the fifth century BC historian Herodotus, Hesiod's retelling of the old stories became the generally accepted version.

Against the mythological background provided by Homer and Hesiod we can place a little political history. Around 1000 BC, Greek colonists, driven from the mainland by the Dorians, established colonies in Ionia, on the west coast of what is now Turkey. In spite of several invasions, which culminated in occupation by the Persians in 546 BC, their twelve cities remained prosperous. These included Miletus, Samos, Ephesus and Colophon, places that some rather well-known people came from or went to, including St. Paul. The Ionian revolt against King Darius I in 500 BC precipitated the Persian Wars. While all this was going on, Greek philosophy is traditionally supposed to have begun in 585 BC when Thales of Miletus correctly forecast an eclipse of the sun. What makes the Greek philosophy of this pre-Socratic period so striking, however, is not that people were suddenly able to perform such mathematical feats as predicting eclipses. The Babylonian astronomers had been doing that for centuries, but they had worked for the most part with purely mathematical relations, like the rhythms of the solar system, whereas the Greeks got into the deepest philosophical questions: How and why did the universe begin? What is it made of and how does it work?[2]

Old farmer Hesiod had answered some of these questions before, and had done so in terms of the creative deeds of gods, but by his time the divine world, formerly transparent to human vision, had become less accessible and many of the stories appear to have reached us in corrupt forms. It is not surprising, therefore, that some of the activities of the Olympian Pantheon seem more human than godlike. This is not merely a modern reaction to the stories – Xenophanes of Colophon (c. 570 – c. 475 BC) complained that "Homer and Hesiod attributed to the gods all the things that among men are regarded as shameful and blameworthy – theft and adultery and mutual deception," and charged the authors with impiety. But he might himself have been accused of impiety, for he actually rejected the whole pantheon of anthropomorphic gods in favor of a single great god who perceives and works through the sheer power of thought.

Xenophanes was not alone in dismissing the Olympian deities in favor of something less personal, more predictable and more logical. Although they differed radically in their interpretations of the structure of the world, the philosophers of the 6th and 5th centuries BC generally agreed that *its processes are lawful and can be understood*. History is not a one-damn-thing-after-another sequence of unrelated events and nature is not just a playground for a troupe of whimsical gods. If you are shocked by these references to the Olympians you should remember that I am reporting Greek opinions, not giving my own. In Plato's *Phaedrus*, Socrates scornfully dismisses a whole range of mythological tales. This was a stage of evolution in which the ancient stories were passed along by people who had largely lost contact with spiritual realities that remained accessible only through the mystery schools.

*

Historians often give us the impression that every historical event, artistic movement and individual achievement is motivated and influenced by preceding events and personalities and is, in that sense, a continuation of something that has been going on already. And yet it seems that around 600 BC, more than a century before the birth of Socrates, this little Greek platoon of natural philosophers sprang deeply motivated, if not fully armed, out of the grass, full of questions regarding the driving force and principles at work in the world.

Aristotle was fully aware of the situation, making, as Jonathan Barnes puts it, "a sharp distinction between what he called the 'mythologists' and the philosophers; and it is true that the differences are far more marked and far more significant than the similarities... It would be silly to claim that the pre-Socratics began something totally novel and entirely unprecedented in the history of human intellectual endeavor. But it remains true that the best researches of scholarship have produced remarkably little by way of true antecedents. It is reasonable to conclude that Miletus in the early sixth century BC saw the birth of science and philosophy."[3]

Barnes ascribes this efflorescence of rational enquiry to "genius" rather than "supernatural talent." While not denying the possibility of genius – whatever that much abused word may mean – we must point out that the key to this remarkable transformation was given by Rudolf Steiner.

*

Steiner's picture of human evolution is in many ways the exact opposite of Darwin's. In Darwin's theory and all its progeny there is nothing purposive; the apparently miraculous organization of even the simplest organisms is the result of statistical inevitability and the self-replicative properties of certain molecular structures. The stages of development reached by present generations of people, animals and plants are finely tuned to conditions of earth and sky, but any appearance of purpose can be traced back to the effects of natural selection and the properties of particles produced by the Big Bang. In this view, the human race was a very late arrival on the scene whereas, according to Steiner, human beings were present in a remote age of the world at the very beginning of evolution, long before our present states of matter appeared. The whole process has been one of physical densification and evolving consciousness and has proceeded under the guidance of successive levels of spiritual beings, who have gone through their own parallel stages of evolution and have appeared to humanity as gods and angels.

Early Christians, encouraged by both esoteric tradition and the words of St. Paul, pictured nine orders grouped in three choirs: Seraphim, Cherubim and Thrones; Dominations, Virtues and Powers; and Principalities, Archangels and Angels. Steiner speaks of these divine powers as having been a constant, active and perceived presence in the creation and evolution of the world and of humanity. It is not quite correct to say that until fairly recent times human beings were able to observe the spiritual powers at work. The word "observe" implies a degree of separation, whereas for a long time none existed. It was the destiny of the human race, however, to eventually achieve a degree of independence that would allow interaction with the spiritual world to take place in freedom. The intention was that we should take hold of life on earth in a new, practical and thoughtful way while keeping some vision and understanding of our spiritual nature. As the spiritual powers withdrew, the early philosophers embarked on the protracted task of replacing divine intimations with independent thought.

*

We are so used to experiencing our thinking as a process that we control, and our thoughts as our own, even when acquired from someone else, that it is hard not to assume that this has always been so. Business records preserved from ancient Egypt seem to have been compiled out of a mercantile disposition not so very different from that of nineteenth century England. From the Babylonian astronomers to the Greek mathematicians, the practical handling of number and geometrical form was conducted in ways that appeal very much to the modern consciousness, and it is easy to overlook the metaphysical excursions and spiritual intimations of those whose perceptions took them beyond the transactional world. And yet the whole flavor of the ancient civilizations is quite removed from anything that we experience today. The objects of everyday experience were physical, certainly, but not merely physical. Thoughts were not merely about the objects of perception but part of the perceptual process. In the ancient Greek mystery schools, as Steiner tells us, there was still some consciousness of the spiritual beings who had stewardship of the workings of nature, and he gives the following description of their experience.

> *I turn my spiritual sight up toward those beings who, through the science of the mysteries, have been revealed to me as the beings of form [Exusiai]. They are the bearers of cosmic intelligence; they are the bearers of cosmic thoughts. They let thoughts stream through all the world events, and they bestow these human thoughts upon the soul so that it can experience them consciously.*[4]

(Exusiai is another name for Powers, the fourth hierarchy above the human being.)

Steiner describes how, in a process centered in the fourth century AD and reaching completion in the fourteenth, the Exusiai gave up their rulership of the cosmic intelligence to the Archai – the Principalities – one step closer to the human being. At the same time, the Exusiai maintained their stewardship of the whole world of sense impressions – colors, forms and sounds. The ancient Greek had perceived the angelic thought-forms streaming from natural objects. During the time of which we are speaking this capacity gradually disappeared, as did the ability to see into the supersensible world. Hitherto people had experienced the thoughts and the actions of the hierarchies as part of their perception of the natural world. Now thinking would come to be an inner experience, while sense perceptions would still be felt as something external, a separation that was to create enormous difficulties for the philosophers of the Middle Ages. If we assume that Steiner meant exactly what he said, we must conclude that the change began in the seventh century BC and extended over the period that he called the Age of the Intellectual Soul, or the Fourth Post-Atlantean

Age. Its completion accounts for the growing feelings of independence and self-confidence with which people in the Renaissance tackled the problems of the world around them. The timing of its inception helps us to understand the impulse to develop objective and logical explanations for the phenomena of the natural world, which appeared in Greek civilization around 600 BC, two centuries before the Golden Age of Greek philosophy.

When we look at some of the work of the philosophers of the fifth and sixth centuries BC, and the difficulties they got into, we shall soon see that it would be a mistake to assume that their thought processes were similar to those of present day scientists. They were living through the earliest stages of a great transition in which the gradual emergence of a new, literally more down-to-earth relationship to thinking did not abruptly terminate their sense of the divine.

Without the advantages of the systematic logic later developed by Aristotle, the pre-Socratic philosophers labored to find the principles of unity that they thought must support the diversity of the phenomenal world. No complete work by any one of them has come down to us, and if it were not for the fact that many later authors quoted them, sometimes at considerable length, we should know hardly anything of them. The quotations and paraphrases that have survived may not always be reliable, but one thing that seems certain is that the original authors contradicted each other, and sometimes themselves, quite freely. It is therefore not easy to grasp the unity of purpose that lay behind their apparently disparate efforts. Thales thought this, Anaximander thought that and Empedocles thought several other things. Parmenides and Heraclitus differed fundamentally about the nature of being, and the resolution propounded by Leucippus and Democritus was unacceptable to most of their colleagues. What they all wanted was to understand.

(ii)
Elements and Principles

To say that nature is self-explanatory means to account for its enormous variety in terms of a few basic principles for which reasons can be given, and here we touch on the Greek idea of the elements. Thales appears to have believed that all the different substances in the universe are transformations of water. His world remained highly numinous. "Some say that soul is mixed with everything," says Aristotle, and "perhaps that is why Thales thought that the universe is full of gods." Anaximander (c. 612 – 545 BC), another Milesian, spoke of an unfamiliar basic stuff, infinite and indefinite, from which all familiar substances derive. Simplicius,[5] writing more than a millennium after the death of Anaximander, reports his belief that "the things from which existing things come into being are also the things into which they are destroyed, in accordance with what must be." This idea makes perfect sense to us. Adam is made from the dust of the earth and to dust he returns. But when Anaximander goes on to say, "for they give justice and reparation to one another in accordance with the arrangement of time," it becomes clear that his thoughts are moving in a different world from that of a modern scientist. Simplicius cautions that "he speaks of them in this way in somewhat poetical words," but Simplicius was writing a thousand years later and may have found it hard to imagine the degree to which Anaximander's primary stuff and its metamorphoses were still ensouled.

Anaximenes, Anaximander's younger contemporary and pupil, thought the infinite stuff was air. Condensation generates winds, clouds, rain, hail and earth. Rarefaction yields fire. These processes bring hot and cold into being. As Plutarch reports, "Anaximenes asserted that… everything comes into being from air and is resolved into it again. Our souls, he says, being air, hold us together, and breath and air contain the whole world." Fifty years or so later, Alcmaeon, the physician, wanted to explain everything in terms of pairs of opposites: hot and cold, light and dark, wet and dry. Xenophanes seems to have taken earth as the primal substance and Heraclitus (fl. 500 BC), the 'obscure' and the 'riddler', who came from Ephesus, made fire his principle. Diogenes Laertius, writing about 200 AD, reports that "Euripides gave Socrates a copy of Heraclitus' book and asked him what he thought of it. He replied: 'What I understand is splendid; and I think that what I don't understand is too – but it would take a Delian diver to get to the bottom of it.'" Hippolytus, a much later historian, gives us a sample: "Heraclitus says that the universe is divisible and indivisible, generated and ungenerated, mortal and immortal, Word and Eternity, Father and Son, God and Justice. He praises and admires the unseen part of his [God's] power above the known part. That he is visible to men and not undiscoverable he says in the following words: 'I honor more those things which are learned by sight and hearing…'" No wonder Socrates was baffled, and Parmenides, like Averroes[6] and Aquinas more than a millennium and a half later, felt it necessary to state quite forcefully that contraries could not be simultaneously true. Modern readers may well feel that they have even more reason for bafflement than Socrates. It is generally agreed, however, that Heraclitus believed that the world is in a state of continuous change and that it is possible to use one's senses and intelligence to understand the way things work, although most people do not do so.

Empedocles (c.495 – 435 BC), who, like Alcmaeon, hailed from a Greek colony in Sicily, imposed some order on this mass of conflicting opinions. Whereas previous writers had generally believed that the multiplicity of nature resulted from transformations of a single basic stuff – water, air, earth, fire, or some uncharacterized substance – Empedocles attributed variety to combinations of earth, water, air and fire, characterized in terms of Alcmaeon's opposites – hot and cold, and wet and dry. This system was accepted by many later philosophers and is what most people think of when "the Greek elements" are mentioned. The elements in different substances, and in the world as a whole, are held together by love and torn apart by strife. To say, as some have, that love simply means a force of attraction and strife a force of repulsion, is to assume naïvely that the ancient Greeks thought in exactly the same way as the nineteenth century scientists but had odd, fanciful ways of expressing themselves. This is a very lazy way of proceeding and has the advantage of dodging the difficult task of understanding what was really in the minds of these writers. Love transforms the elements, just as it transforms men and women. The modern concepts of element and atom are inseparable, but Empedocles paid very little attention to atoms. He believed that truth is to some extent attainable and can be grasped as far as human reason reaches, and he attacked those who claimed to know more, in terms that make it clear that he lived in a world in which the development of an independent human consciousness had only just begun.

But, O Gods, turn the madness of these men from my tongue,
And from holy mouths channel forth a pure spring.
And you, Muse of long memory, white-armed maiden,
I beseech: what is right for mortals to hear,
Send to me, driving the well-reined chariot of piety.

Evil men want to have power over truth by distrusting it and try to gain reputation and honor by pretending to greater knowledge than is possible or right, but the Muse assures Empedocles that an understanding of the processes of the natural world is open to those who have the right relationship to divine wisdom.

Happy is he who has gained a wealth of divine thoughts, wretched he whose beliefs about the gods are dark.

In other words, it was permissible for people to investigate nature but they still couldn't quite be trusted to do it on their own. By the time of Empedocles, however, a great difficulty had arisen which had somehow to be cleared up. Xenophanes had already voiced scepticism about the possibilities of physical science, and such doubts were stated more emphatically by three philosophers usually associated with Elea, a settlement on the west coast of what is now southern Italy, although one of them lived on the island of Samos in the Aegean Sea.

(iii)
The Way of Truth

Parmenides, who flourished in the early fifth century BC, was the chief hero (or villain) of this new movement. He composed a long poem, of which the second half, *The Way of Opinion*, was a description of nature. In the first part of the poem, however, he had already made it clear that *The Way of Opinion* was false and misleading. *The Way of Truth*, as the first part is known, is a difficult and highly influential document. It is fortunate that thanks to the efforts of later commentators, philosophers, and historians, a reasonably reliable text for the first part can be assembled.

The Atom - A Historical Background—13

One thing that makes *The Way of Truth* difficult for a modern reader is the apparently fanciful imagery through which the thought is conveyed. One can equally well say, however, that the imagery is true to the actual cosmic situation. The Goddess gives Parmenides guidance and encouragement and tells him that *from this point on he must begin to take responsibility for his own thinking.* Guided by the daughters of the sun, Parmenides sets out "on the celebrated road of the god which carries the man of knowledge..." They reach the doors of night and day where, appeased with the soft words of the maidens, the Goddess Justice receives Parmenides. "You must learn all things," she says, "both the steadfast heart of compelling truth and the untrustworthy opinions of mortal men."

The poem continues:

But come, I will tell you
The only roads there are to be thought of:
One that is and cannot not be, is the path of reason, (for
    truth accompanies it);
Another, that is not and must not be,
A trail devoid of all knowledge.
For you could not recognize that which is not (for it is
    not to be done),
Nor could you mention it.

The road "that is, and cannot not be" is the path of persuasion, reason, thinking. For our purposes, the most important aspect of the road "that is not and must not be," the "trail devoid of knowledge," has to do with the need that some philosophers felt to include both being and non-being in their world view on an equal footing. As we shall see, the existence of non-existence – in other words, void or pure emptiness – was an essential part of Leucippus' and Democritus' atomic theory, and this is what Parmenides emphatically rejects. Along this false road,

> Mortals who know nothing wander, two-headed...
> And they are borne along
> alike deaf and blind, amazed, undiscerning crowds,
> for whom **to be and not to be** are deemed the same
> and not the same.

It is impossible to refrain from commenting that this may account for Hamlet's difficulty in making up his mind, for Lear's remark that "Nothing can come of nothing," and for the Irish lady who offered me two kinds of ice cream that were "the same only different."

According to Plato, Socrates had met Parmenides:

> When we were boys, my boy, the great Parmenides would testify against this [namely the view that what is not is] [7] from beginning to end, constantly saying, 'Never will this prevail, that what is not is: restrain your thought from this road of enquiry.'

Parmenides' rejection of non-being as a participant in the processes of the world and belief in the priority of thinking over appearances lead to the conclusion that in spite of all the appearances, creation and dissolution are impossible and the world must be regarded as a single, eternal, unchangeable object.

Melissus of Samos and Zeno of Elea, the two philosophers usually grouped with Parmenides as members of the Eleatic School, adopted this view of the cosmos. The little that is known of Melissus suggests that he was quite a character. The only definite date connected with him is 441BC, when the island of Samos was attacked by the Athenians. According to Plutarch, Melissus persuaded his fellow citizens to take advantage of the temporary absence of the Athenian leader, Pericles, and some of his ships. They launched a successful attack, in which the Samians "captured many men and destroyed many ships, thereby gaining control of the sea and acquiring many supplies for the prosecution of the war.... Aristotle says that Pericles himself had earlier been defeated by Melissus in a sea-battle." In the long run the Athenians prevailed, but, as Jonathan Barnes remarks, "Melissus had made a mark on history unusual for a philosopher."[8]

Melissus made his mark on philosophy by putting the essence of Parmenides' obscure poem into plain prose and adding some thoughts of his own. Here are some samples, preserved by the admirable Simplicius:

Whatever existed always existed and always will exist. For if it had come into being, then necessarily before coming into being it would have been nothing. Now if it had been nothing it would in no way have come to be anything from being nothing.[9]

If Parmenides and Melissus are correct, science is impossible. Sense impressions are not to be trusted and the appearance of change is an illusion.

[The cosmos] is eternal and infinite and one and wholly homogeneous. And it will neither perish nor grow larger nor change its arrangement nor suffer pain nor suffer anguish. For if it underwent any of these things it would no longer be one.

...Nor is it empty in any respect. For what is empty is nothing: and being so it would not exist.

Finally, speaking of the apparent multiplicity of objects:

...they all seem to us to alter and to change from what they

were each time they were seen. So it is clear that we do not see correctly, and that those many things do not seem correctly to exist. For they would not change if they were true, but each would be as it seemed to be: for nothing is stronger than what is true...

Nothing is stronger than what is true. This could be taken as a motto for the whole Greek philosophical endeavor. The belief that soundly argued philosophical conclusions are compulsory, whatever the appearances, continued for many centuries, in spite of the contradictions which frequently arose between different schools and individuals. Truth, it seems, wears many faces.

*

It would be only too easy to pick out the elements of apparently modern thought and ignore the spiritual background out of which all this thinking arises. There are some useful observations about states of matter, for instance, and Parmenides was aware that the earth is round. Viewed with modern hindsight, however, most of the scientific observations seem to be nonsense. The real point about the Pre-Socratics is not that their science was correct but that they arrived at their conclusions by applying the power of thought to common observation. Jonathan Barnes praises them for inventing the very idea of science and philosophy, and for providing the beginnings of a vocabulary. "What, then", he asks, "is the substance of the claim that the Pre-Socratics were the champions of reason and rationality? It is this: they offered reasons for their opinions, they gave arguments for their views.... Perhaps that seems an unremarkable achievement. It is not.... Those who doubt the fact should reflect on the maxim of George Berkeley, the eighteenth century Irish philosopher: All men have opinions but few think."

In the fifth century BC, however, it was thinking that seemed to be about to bring the evolution of science to a halt, for science depends precisely on observing the changes that the Eleatic philosophers considered illusory. As Rudolf Steiner puts it, it was as if Parmenides were surrounded by a wall of thought,[10] cutting him off from the world of natural phenomena, with the result that his view of the world was in direct opposition to that of Heraclitus. This schism was the direct result of the decision of the hierarchies to let human beings separate their thinking from their sensory experience of nature and it led to the early formulation of the atomic theory.

(iv)

Atoms

Expressed in modern terms, the problem was how to reconcile the inner need for permanence with the outer perception that everything is always changing, and it seemed for a while that the solution lay in the atomic theory advanced by Leucippus, a mysterious character whose dates and place of birth are uncertain, whose writings are almost entirely lost, and who seems to have "flourished" about 460 BC.

More is known about Democritus than about Leucippus. Only fragments of his work survive, but he was extensively quoted by later writers, including Aristotle. He is known to have been born in Abdera, in the north of Greece, probably around 460 BC, and is said to have traveled widely and to have learnt from Leucippus. He was a hugely prolific author on atomism, science, literature, the nature of knowledge, and ethics. His philosophy of atomism was preserved by the Epicureans and became influential in Western European thought. Aristotle's essay on Democritus is lost, but Simplicius quotes a useful fragment from it. After bringing out the idea of an infinite number of permanent particles situated in an infinite void, Aristotle continues:

> He [Democritus] thinks that the particles are so small that they escape our senses and that they possess all sorts of forms and all sorts of shapes and differences in magnitude. From them, as from elements, he was able to generate and compound visible and perceptible bodies. The atoms struggle and are carried about in the void because of their dissimilarities and the other differences mentioned, and as they are carried about they collide and are bound together in a binding which makes them touch and be contiguous with one another, **but which does not genuinely produce any other single nature whatever from them**; for it is utterly silly to think that two or more things could ever become one. He explains how the particles remain together in terms of the ways in which they entangle with and grasp hold of one another, for some of them are uneven, some hooked, some concave, some convex, and others have innumerable other differences. So he thinks that they hold on to one another and remain together until some stronger force reaches them from their environment and shakes them and scatters them apart. He speaks of generation, and of its contrary, dissolution, not only in connection with animals but also in connection with plants and worlds – and in general with all perceptible bodies.

The phrase which I have placed in bold letters gives some justification for placing Democritus in the camp of Parmenides, rather than that of Heraclitus. The appearance of change is genuine, but it *is* only an appearance.

It seems clear from internal evidence, and is generally believed, that atomism is an attempt to reconcile the apparently irreconcilable contradiction between an immovable philosophical object, permanence, and an irresistible natural force, change; between the ineluctable results of thinking and the overwhelming evidence of the senses. What is not clear is that Democritus ever gave this explicitly as a reason for his belief in the theory of atoms and void. Aristotle supplies a different argument, based on the paradoxes which would arise if matter were infinitely divisible. This is a justification for the word *atom* – that which cannot be divided. The fact is, however, that the major impact of the atomic theory has always been on the struggle to understand the mechanisms of change in relation to an underlying concept of permanence. Atoms have to be indivisible; otherwise they would not provide the basis of permanence. They have to move about in order to provide the appearance of change. The primal substance of which the atoms are made has no perceptible properties and is never transformed from one nature to another. Atoms never change from one form to another. Their motions produce different configurations, which are not permanent but are lasting enough to create the semblance of different kinds of matter. If we were able to see the individual atoms in their patterns we should see that no fundamental change takes place at all.

This all seems to be very simple, elegant and effective, but there are some serious problems, the first of which is that we have to ignore, or somehow get around, Parmenides' strenuous objections to allowing non-being – void – into the system. Democritus seems simply to have asserted the paradoxical existence of non-being as an essential part of the system, an idea which was anathema not only to Parmenides but also to later thinkers, such as Descartes and Robert Boyle, and which raises some fascinating questions in relation to modern physics.

Another problem concerns the motion of the atoms; "As for motion," says Aristotle, "whence and how existing things acquire it, they [Leucippus and Democritus] negligently omitted to inquire about it." They also seem to have omitted to inquire where the atoms, with or without motion, came from. Aristotle was committed to the idea of the prime mover. Whatever is changing is moving from potentiality to actuality. Objects change to a new state through contact with that which is already in the new state – as when objects are set in motion by things that are already in motion or made hot by contact with things that are already hot. But what of the change before changes when everything was new? Aristotle believed that it was logically necessary that the first motions of all were brought about by a prime mover which must itself be motionless. Theologians of all stripes identified the prime mover with God. We can only assume that Democritus thought that motion, like shape and magnitude, is a primitive property of the atoms, rather than the gift of a divine power, and that the universe is without beginning or end. It seems hardly likely, at any rate, that such questions would have escaped his notice.

(v)

Roadblock

Perhaps the most distressing difficulty concerns the question of whether it is really possible to know anything at all. According to Sextus Empiricus, who flourished in the third century AD and seems to have been the greatest living authority on Scepticism:

> Democritus sometimes does away with what appears to the senses and says that nothing of this sort appears in truth but only in opinion, truth among the things that exist lying in the fact that there are atoms and void. For he says, 'By convention sweet and by convention bitter, by convention

> hot and by convention cold, by convention color: in reality atoms and void.' That is to say, objects of perception are thought and believed to exist but they do not exist in truth – only atoms and void do. [Democritus also says] '... in reality we know nothing about anything...' Now in these passages he does away in effect with all knowledge, even if it is only the senses which he explicitly attacks. But in the Rules he says that there are two forms of knowledge, one by way of the senses and the other by way of the understanding. The one by way of the understanding he calls genuine, and deems it reliable as a way of finding out what is true; the one by way of the senses he calls dark, denying that it leads unerringly to the truth.

So, by way of an atomic theory which seems to have been designed to overcome a philosophical *impasse*, Democritus leads us back more or less to square one and places us firmly in the arms of Parmenides. "Nothing is stronger than what is true." We arrive at what is true by means of reason, ignoring the deceptive claims of our senses. Democritus' view that sense impressions are purely conventional and that the reality is atoms and void was repeated by several commentators and led to the general impression that Democritus believed that knowledge is impossible. The problem was stated briefly by Galen the physician, who lived in the second century A. D. and was deeply interested in the philosophy of science:

> Everyone knows that the greatest charge against any argument is that it conflicts with what is evident. For arguments cannot even start without evidence: how then can they be credible if they attack that from which they took their beginnings? Democritus too was aware of this; for when he had brought charges against the senses, saying, 'By convention color, by convention sweet, by convention bitter;

> *in reality atoms and void,'* he had the senses reply to the intellect as follows: *'Poor mind, do you take your evidence from us and then try to overthrow us? Our overthrow is your fall.'*
>
> So one should condemn the unreliability of an argument which is so bad that its most persuasive part conflicts with the evident propositions from which it took its start.

This type of cyclical argument has often been repeated and is stated most persuasively in J. B. S. Haldane's *Possible Worlds*[11] and quoted by C. S. Lewis in *Miracles*:

> *If my mental processes are determined wholly by the motions of atoms in my brain, I have no reason to suppose that my beliefs are true... and hence I have no reason for supposing my brain to be composed of atoms.*[12]

The atomic theory had virtually no impact on scientific thinking until the late seventeenth century, and we might be tempted to conclude that it is this logical roadblock that accounts for its long stay in limbo. It seems, however, that those who worry about such things are always in the minority and have little immediate influence. If the system seems to work well enough from a practical point of view the majority will want to get on with it, ignoring, or unaware of, the philosophical or ethical qualms of the minority.

*

The ability to see, to some extent, into the spiritual world waned slowly, and the great Pagan and Christian syntheses of thinking and revelation – Neoplatonism and Thomism – were only gradually displaced by a nominalist view which provided fertile soil for the growth of materialistic theories. Even after the end of the Fourth Post-Atlantean Age, Francis Bacon, who is popularly supposed to have been responsible for the materialistic, industrial, technological trend of modern civilization, had little interest in atomic theories. While we cannot do more than scratch the surface of later Greek philosophy, we can at least get an impression of the ways in which the need for such a theory was obviated.

The Greek philosophers and their heirs were not automatically against atomic theories, but they were able to treat the problems which had given rise to the atomic theory in a manner more in harmony with the spirit of the time. Plato's excursions into atomic theory are real enough, but they do not lead to a materialistic view and are of minor importance in his work as a whole. His world of Ideas projecting forms into the physical world gave validity to the possibilities of ideal permanence and physical change.

According to Aristotle, Parmenides erred insofar as he did not take into consideration the different meanings of *being*. A tree *is*; the height of the tree *is*. The being of the tree is substantial; the being of its height is accidental. Accidental being is distinct, but not separate, from substantial being. Furthermore, the acorn is not an oak in substance but it is an oak in potential. Aristotle thus distinguishes between *being in potential* and *being in act*. That which did not exist in act can come into existence through the fulfillment of potential. This is what we may call *substantial change*. *Accidental change* takes place when an object remains essentially itself while its physical dimensions change. Aristotle goes into these distinctions in great depth and detail and makes the atomic theory appear philosophically unnecessary. In ancient and medieval times Nature was more a world to be lived in and lived with, and less a field for manipulation and exploitation than it became later. Natural organic processes did not need mechanistic explanations. Democritus provided a theory designed to overcome a philosophical difficulty, but it raised new problems, gave only a limited, general explanation of well-known phenomena and made no attempt to predict new ones.

It was, in fact, a theory that most of the philosophers could well do without. Having been proposed soon after the beginning of the period in which, through the activity of spiritual powers, human beings were gradually to take responsibility for their thinking, it could not incarnate effectively into the world until this process was complete. The theory did not reappear punctually at the end of the fourteenth century because a new relationship to nature had to be developed before it could be used effectively and because the soul capacities which had slowly developed over a period of two thousand years were and are still in the process of being mastered. The stirrings of independence in the realm of thinking, evident among the Pre-Socratics, continued to be restrained by traditions and authorities well into the Renaissance. To this day appeals to authority, rank and pedigree have not vanished from the earth among scientists who become enamored of position, priority and their own theories, and even among anthroposophists, from whom one might have expected particular devotion to the ideal of spiritual freedom.

During the long period between Democritus (460-370 BC) and Newton (1642-1727 AD) the atomic theory existed largely in potential. The Epicureans, who took it up later in the fourth century BC, found that it provided a good philosophical background for their atheistic hedonism, and the most extensive ancient writings that we have on it come from the Roman Epicurean, Lucretius (99-55 BC). In the absence of any original texts from Democritus it is hard to tell how much of the extensive philosophical speculation on the activities of atoms indulged in by the Epicureans was original and how much mere repetition. What is clear is that there was no essential change in the basic physical concept of atoms moving in the void. In any case Epicureanism was much more influential as the basis of a life style than as a serious philosophy. The important thing is that it provided a channel through which atomism could pass into the idea pool of the Renaissance.

(vi)
Atoms back in Vogue

I referred earlier to the Fourth Post-Atlantean Age, also known as the Age of the Intellectual Soul, during which human beings were to develop freedom in the realm of thinking. The succeeding age, that of the *consciousness soul*, began early in the fifteenth century, and is the period in which we have the potential to use our freedom to begin the process of working back into full and conscious communion with the spiritual world. As free human beings we experience our individuality and separateness, so that we are in danger of being overwhelmed by a feeling of isolation from both the spiritual and the physical world. To be at the same time both separate and involved – in other words, to interact without losing one's individuality – demands a fine balance between thinking and feeling that is not always easy to achieve. Atoms neither think nor feel, but in the consciousness soul age they did acquire a problem analogous to that of the human being.

The atoms of Democritus were indestructible and unchangeable, and could interact only by physical contact.[13] During the nineteenth century it became clear that atomic interactions were complex and some kind of compromise between permanence and changeability would have to be worked out. Like the human being, atoms had to be both separate and involved.

René Descartes (1596-1650) was one of several thinkers who brought the atom back to popularity in the seventeenth century. Scientists of the seventeenth and subsequent centuries did not like the idea of empty space any better than Parmenides, so they did their best to provide their atoms with some sort of medium to move around in, such as a kind of cosmic dust produced by originally cubical atoms grinding together until they became spherical.

One necessity for the inception of the modern phase was the metamorphosis of the Greek and, later, the Paracelsian notion of the element into something approaching the modern idea. Although neither the Greeks' earth, water, air and fire, nor the Paracelsian quicksilver, sulphur and salt were expected to function in the same way as the chemical elements of the nineteenth century, as soon as chemistry became an actual experimental science it seemed necessary to refute the ancient ideas as if such had been their intention. Robert Boyle (1627-1691), who dealt with the older notions in *The Sceptical Chymist*, used atomic concepts as a help in clarifying his ideas on elements and compounds. Isaac Newton (1642-1727) took the major step of combining mathematical principles and atomic concepts in an effort to account for the behavior of bulk matter, and doesn't seem to have been bothered by the use of "non-being" or vacuum in the theory.[14] In the last part of his *Opticks*, first published in 1704, he voices his approval of "the oldest and most celebrated Philosophers of Greece and Phoenicia, who made a Vacuum and Atoms and the Gravity of atoms the first Principles of their Philosophy." However, unlike Democritus, he did not "negligently omit to enquire into their origin," but gave all the credit to God.

... it seems probable to me that God in the beginning formed Matter in solid, massy, hard, impenetrable, moveable Particles, of such Sizes and Figures, and with such other Properties, and in such Proportion to Space, as most conduced to the End for which he form'd them; and that these primitive Particles being Solids, are incomparably harder than any porous Bodies compounded of them; even so very hard, as never to wear or break in pieces; no ordinary Power being able to divide what God himself made one in the first Creation. While the Particles continue entire, they compose Bodies of one and the same Nature and Texture in all ages: But should they wear away or break in pieces, the Nature of Things depending on them would be changed.... And therefore, that Nature may be lasting, the Changes of corporeal Things are to be placed only in the various Separations and new Associations and Motions of these permanent Particles...[15]

In explaining how the combination of atomic theory and the mechanical laws of motion should be used to explain the nature and operation of the material world, Newton recognizes the need for some principles of interaction.

It seems to me farther, that these Particles have not only a Vis Inertiae [force of inertia], *accompanied with such passive Laws of Motion as naturally result from that Force, but also that they are moved by certain active Principles, such as is that of Gravity, and that which causes Fermentation* [breaking down], *and the Cohesion of bodies. These Principles I consider, not as occult* [hidden] *Qualities, supposed to result from the specifick Forms of Things, but as general Laws of Nature, by which the things themselves are formed; their Truth appearing to us by Phaenomena, though their Causes be not yet discover'd.... To tell us that every Species of Things is endow'd with an occult specifick Quality by which it acts and produces manifest Effects, is to tell us nothing:* **But to derive two or three general Principles**

> *of Motion from Phaenomena, and afterwards to tell us how the Properties and Actions of all corporeal Things follow from these manifest Principles, would be a very great step in Philosophy, tho' the Causes of those Principles were not yet discovered.*

If, at first sight, it appears that Newton's atomism has progressed beyond that of Democritus only to the extent that he has given it a theological justification, we must go on to see that he believes that natural phenomena ought to be quantitatively explicable in terms of atoms. We know the laws that govern the motions of ordinary terrestrial objects and the heavenly bodies of the solar system. If we apply these laws to the motions of the atoms we should be able to understand how they lead to "the Properties and Actions of all corporeal Things."

To put it baldly: *All of material nature should ultimately be reducible to atomic mechanics.*

This doesn't rule out the presence of spirit but it does imply a strict separation between the material and the spiritual.

*

Newton provided the link between the Democritan theory, which might be regarded as an unfortunate philosophical dead end, and the nineteenth century theory, in which the atoms are subject to Newtonian mechanics and held accountable for the discernible properties of matter. The earliest fruits of this approach ripened in Proposition 23 of the *Principia* of 1686, where Newton calculated the properties of a fluid[16] composed of atoms with certain specified characteristics and found that such a fluid would conform exactly to the recently observed relationship between the pressure and volume of a gas, now known as Boyle's Law.[17]

He properly pointed out that one must not conclude from this that actual gases are composed of such atoms – which is just as well since, a few decades later, Daniel Bernoulli got precisely the same result with a totally different set of assumptions.

Newton had created what became known as a *conceptual model* of a fluid. He explained the outward pressure of the gas by assuming that there are forces of repulsion between the atoms. His atoms were more or less stationary and the gas acted like a three-dimensional spring. Bernoulli's atoms, however, moved freely about and didn't interact with one another at all, the outward pressure being caused by their impacts on the walls of the containing vessel.

We must note that Newton, who was a great admirer of Boyle and his work, *does not mention air or any known gas* and does not refer to Boyle or any experimental evidence that supports the relationship between pressure and volume which Boyle had discovered. Newton uses the word "fluid," and the two theorems preceding this one in the *Principia* deal with fluids that conform to Boyle's Law. Looking back at the previous proposition, we find some clues. "Let the density of any fluid be proportional to the compression" (an alternative way of stating Boyle's Law) he says in Proposition 22, and goes on to derive what is recognizable as a correct account of the way in which the density of the actual air in the earth's atmosphere decreases with altitude. But to a physicist "fluid" means anything that flows – liquid, vapor or gas.[18] *Real* liquids and vapors do not "obey" Boyle's Law. Reading Newton's analysis we might thoughtlessly assume that he was talking about actual gases, whereas he was really speaking of a purely hypothetical fluid. Newton was perfectly straightforward on this point. In the commentary on Proposition 22 he remarks that "as to our own air, this is certain from experiment, that the density is either accurately, or very nearly at least, as the compressing force" – meaning that Boyle's Law is at least a good approximation to the truth – and is pretty sure that his conclusion indeed applies to the earth's atmosphere. This makes sense, since it is worked out without any assumptions about atoms and their properties, and gives a verifiable result. About Proposition 23 he is noncommittal. "But whether elastic fluids do really consist of particles so repelling each other, is a physical [not mathematical] question. We have here demonstrated mathematically the property of fluids consisting of particles of this kind, that hence [natural] philosophers may take occasion to discuss that question." Later writers were not always so frank about their speculations.

Bernoulli's idea was the germ of what became known as the Kinetic Theory of Matter, which was developed throughout the nineteenth century and had considerable success in explaining the properties of gases but didn't make much of a dent in the understanding of solids and liquids. More sophisticated forms of the theory, such as that of van der Waals, involved the assumption, contrary to Newton and Bernoulli, that there were forces of *attraction* between the atoms, resulting in a form of the gas laws that at least suggested the possibility that there might be such things as liquids and solids. While physicists generally seem to have been cautiously optimistic that they were on the right track with the atomic theory, things were different among the chemists.

*

The present-day concepts of elements, mixtures and compounds and the nature of combustion are so well known to students of elementary chemistry that it often comes as a shock to learn that in 1800 they were still struggling for acceptance. The key figure in the struggle was Antoine Lavoisier[19] (1743-1794), who waged virtually a one-man battle in favor of a new theory of combustion and a new understanding of elements and compounds.

It had always seemed that when a substance burned, something like its vital spirit departed from it and all that was left was ash. This spirit, known as *phlogiston* or fire-stuff, was not fire itself, but its departure gave the appearance of fire. As chemistry gradually became a quantitative science, starting, perhaps, with van Helmont (1579-1644) in the early seventeenth century, it became clear that there were problems with the phlogiston theory. After burning 62 lb. of oak charcoal and finding that 1 lb. of ash remained, van Helmont concluded that 61 lb. of *spiritus silvestre* – wild or woody spirit – had escaped with the flames. This is the kind of result that made the phlogiston theory so attractive, but when it was found that the calx or, as we would say now, oxide, left by a burning metal weighs *more* than the original metal, things began to look different. In spite of a determined rearguard action by several eminent chemists, Lavoisier's well-publicized experiments with the red oxide of mercury and his eloquent advocacy of the oxygen theory carried the day. Physical science up to the end of the eighteenth century had made ample use of invisible, intangible fluids or principles as explanatory devices. In addition to phlogiston, there were the aether, the two electrical fluids and the caloric fluid. Some of these persisted well into the nineteenth and even the twentieth century, but by 1810 or so the phlogiston theory had disappeared without leaving much of a trace. Combustion was explained as a process in which the burning substance combines with oxygen out of the air and the first halting steps were taken towards an understanding of chemical elements and compounds in terms of atoms and molecules.

The real breakthrough happened when John Dalton (1766-1844) published his *A New System of Chemical Philosophy* in 1808. Dalton, who came from an impoverished Quaker family, became a teacher at the age of twelve, a farm laborer at thirteen and a college professor of mathematics and natural philosophy, largely self-taught, at twenty-seven. In 1787 he began to keep a diary of weather and atmospheric conditions, and he stuck to it for fifty-seven years, during which time he recorded over 200,000 observations. His *Meteorological Observations and Essays*, published in 1793, was a highly original document containing the germs of some of his later discoveries. He is remembered today for his work on color-blindness, the partial pressures of gases and, principally, the first attempt to put the atomic theory on a sound quantitative basis.

Dalton's laws of chemical composition are well known to every student of elementary chemistry, so it will be sufficient to say here that he found it easy to account for the proportions in which the chemical elements combine by making the following assumptions:

> 1) Matter is composed of atoms – particles which are too small to see individually and can neither be created nor destroyed.
>
> 2) Chemical reactions involve only the union and the separation of atoms.
>
> 3) Each element has its own particular kind of atom, different from the atoms of all other elements. Each chemical substance which is not an element consists

of compound atoms [molecules] peculiar to that substance and different from those of all other substances.

These highly Democritan assumptions accounted for the constant composition of chemical compounds and for such experimental results as the analyses of the two oxides of carbon. For a given weight of carbon, one of the oxides contains exactly twice as much oxygen as the other. This is explained by saying that a molecule of the former contains one atom of carbon and two of oxygen, while a molecule of the latter contains one atom of each element.

Once you have grasped the idea, the whole thing seems so simple that it is easy to forget the momentous nature of the undertaking. Dalton conceived the idea of finding the relative weights of atoms of all the different elements, defining the atomic weight as the ratio of the weight of one atom of an element to that of one atom of hydrogen. What he proposed that was entirely new to the chemists was the transference of the quantitative relationships found in the visible world of the laboratory into the invisible world of the atom. Newton had believed that the mechanics of observable objects must be explainable in terms of the mechanics of invisible atoms, but his efforts in that direction in explaining the properties of fluids and the behavior of light rays had not gone very far. Taking his inspiration from Newton, Dalton planted a seed that after a long period of slow and difficult growth would branch out in many directions, producing fruits that would vary widely in their edibility and leave a bitter taste in many mouths. There were many dissenters, and opposition continued even after the publication of early versions of the Periodic Table in the 1860's, a few years before Steiner's first encounter with atoms. A superficial view of the late nineteenth century suggests that the atomists grew in numbers and confidence and in the hope that at some point it might be possible to explain all phenomena, including those of human consciousness, in terms of atomic motion. The truth is, however, that physicists and chemists had no idea why particular combinations of atoms were formed and what held the atoms together. The scientists were continually plagued with problems, and the partial solutions that they obtained always led to further difficulties.

(vii)

Making Waves

Newton had also made a kind of atomic theory of light, proposing a set of particle types that are responsible for the prismatic spectrum of colors. This idea, which aroused Goethe's fury, implied that the prism merely separated colors that were already present, rather than creating them through a process of metamorphosis. While Goethe was developing his theory of color early in the nineteenth century, Thomas Young and Augustin Fresnel were laying the foundations of a wave theory of light that would eventually displace Newton's corpuscular theory. Several streams of nineteenth century scientific thought came together in the great figure of James Clerk Maxwell (1831-1879), who took up Young's three-receptor theory of color vision, did fundamental work on the kinetic theory of gases and gave the first electromagnetic interpretation of Young's and Fresnel's light waves. The 1870's and '80's were a time of growing excitement among scientists who knew what was going on, and by the time Steiner was a university student he could hardly have been unaware that a new age was dawning for the physical sciences. People often think of this new age as starting twenty years later with the discoveries of radioactivity, the electron and the nucleus and the first formulations of Planck's quantum theory and Einstein's relativity theory. The latter part of the nineteenth century, up to the 1890's, has been portrayed as a period of confidence, but the first stirrings of uneasiness had already appeared in the 1860's when Maxwell encountered inexplicable contradictions between the kinetic theory of gases and experimental observations at low temperatures.

It is very disconcerting to have an analysis that rests unassailably on accepted scientific principles and to find that sometimes it works and sometimes it doesn't. Maxwell described the problem as "the greatest difficulty yet encountered by the molecular theory." In the words of Richard Feynman, "One often hears it said that physicists at the latter part of the nineteenth century thought that they knew all the significant physical laws and that all they had to do was to calculate more decimal places. Someone may have said that once, and others copied it. But a thorough reading of the literature of the time shows they were all worrying about something. Jeans said about this puzzle that it is a very mysterious phenomenon, and it seems as though as the temperature falls, certain kinds of motions 'freeze out.'"[20] Apparently the molecules do some of the things they are supposed to do, but not all of them and not all the time. Forty years elapsed before a plausible reason for this freezing out process appeared, and when it did it was in the form of the quantum theory. That, however, takes us a long way ahead of our story – the atom that Steiner met as a schoolboy was still really quite a simple thing.

(viii)
Rudolf Steiner meets the Atom

In his autobiography Steiner describes how he met the atomic theory head on in 1872 at the age of eleven.

The principal of my school, in one of the annual reports which had to be issued at the close of each school year, published a lecture entitled 'Attraction Considered as an Effect of Motion.' As a child of eleven years I could at first understand almost nothing of the content of this paper; for it began at once with higher mathematics. Yet from some of the sentences I got hold of a certain meaning. There formed itself in my mind a bridge between what I had learned from the priest concerning the creation of the world and these sentences in the paper. The paper referred also to a book which the principal had written, 'The General Motion of Matter as the Fundamental Cause of All the Phenomenon of Nature.'

I saved my money until I was able to buy that book. It now became my aim to learn as quickly as possible everything that might lead me to an understanding of the paper and the book. The thing was like this. The principal held that the conception of forces acting at a distance from the bodies exerting these forces was an unproved "mystical" hypothesis. He wished to explain the "attraction" between the heavenly bodies as well as that between molecules and atoms without reference to such "forces." He said that between any two bodies there are many small bodies in motion. These, moving back and forth, thrust the larger bodies. Likewise these larger bodies are thrust from every direction on the sides turned away from each other. The thrusts on the sides turned away from each other are much more numerous than those in the spaces between the two bodies. It is for this reason that they approach each other. "Attraction" is not any special force, but only an "effect of motion." I came across two sentences stated positively in the first pages of the volume:

> *"1. There exist space and in space motion continuing for a long time.*
> *2. Space and time are continuous and homogeneous, but matter consists of separate particles (atoms)."*

Out of the motions occurring in the manner described between the small and great parts of matter, the professor would derive all physical and chemical occurrences in nature. I had nothing within me which inclined me in any way whatever to accept such a view; but I had the feeling that it would be a very important matter for me when I could understand what was in this manner expressed. And I did everything I could in order to reach that point. Whenever I could get hold of books of mathematics and physics, I seized the opportunity. It was a slow process. I set myself to read the paper over and over again; each time there was some improvement.[21]

Thanks to the efforts of other teachers, Steiner became absorbed in geometry and the laws of probability and, as time went on, in the general study of mathematics and physics. Even as a child he soon came to regard all this as grist to the spiritual mill.

Behind all that I was taking into myself from the principal, the teacher of mathematics and physics, and the teacher of geometrical design, there arose in me in a boyish way of considering the problem of what goes on in nature. My feeling was: I must go to nature in order to win a standing place in the spiritual world, which was there before me, consciously perceived.

I said to myself: 'One can take the right attitude toward the experience of the spiritual world by one's own soul only when one's process of thinking has reached such a form that it can attain to the reality of being which is in natural phenomena.' With such feelings did I pass through life during the third and fourth years of the Realschule. Everything that I learned I so directed as to bring myself nearer to the goal I have indicated.

This insight is reflected in his development of the way of seeing nature that is inherent in Goethe's scientific work and, in a different way, in his insistence, many years later, that "...nobody can attain a true knowledge of the spirit who has not acquired scientific discipline, who has not learned to investigate and think in the laboratories according to the modern scientific method."[22]

(ix)
Rejection

Steiner was entering a scientific milieu in which explanations of physical phenomena were sought in terms of material particles and electromagnetic waves, and whose outer confidence was disturbed by undercurrents of deep anxiety. Like Ernest Rutherford, the discoverer of the nucleus, Steiner grew up in a world in which people were "brought up to look at the atom as a nice hard fellow, red or grey in color, according to taste." In this "billiard ball" atomic physics there was no knowledge of the atoms' internal structure and no one had the faintest idea of how they interacted with electricity and electromagnetic waves. By the 1880's, some highly suggestive work on the conduction of electricity through gases and aqueous solutions was joining the repetitive nature of the Periodic Table in pointing towards an atom that had some internal structure, but Steiner's most pressing concern, as shown in his prefaces to the Goethe Edition (about which I shall have more to say at a later stage) was in the realm of human consciousness and physiology. The effort to explain and connect all phenomena, including physiological sensations, by means of what Steiner saw as an abstract substratum of auxiliary concepts such as atoms, light waves and energy, resulted in a tendency to dismiss those sensations as purely subjective. Steiner found this tendency very objectionable and echoed the misgivings of Democritus and Galen:

> Those whose capacity for conceiving ideas has not been corrupted by Descartes, Locke, Kant and the modern physiologists will never comprehend how light, color, sound and heat can be considered only subjective states within the human organism and yet an objective world of processes outside this organism can be affirmed.[23]

> It is these reflections that compelled me to reject as impossible every theory of nature which, in principle, extends beyond the domain of the perceived world, and to seek in the sense-world the sole object of consideration for natural science.[24]

And so, "The theory must be limited to the perceptible and must seek connections within this."

These remarks from around 1890 make it clear that as a young man Steiner rejected the atomic theory of his time, and the manner of his rejection suggested that he was unlikely to change his mind. It is worth noting that Steiner was by no means alone in rejecting the atomic theory and that the last quotation might easily have come from the lips of Ernst Mach and, a couple of generations later in a different context, Werner Heisenberg. Heisenberg and Mach, however, had different ideas about what was perceptible and their thoughts went in very different directions from Steiner's.

Steiner did not reject contemporary scientific method in toto, as we shall see when we come to the great scientific lecture courses that he gave in the period 1919-23, but he was deeply troubled by the increasingly pervasive influence of everything to do with electricity, both in scientific thinking and in everyday life. Advances in electrical science and engineering were not to be deplored or ignored, however; Steiner's main concern was that we should enter the electrical age in full consciousness and understand its potential for evil.

(x)

The Age of Electricity

As a schoolboy studying chemistry around 1949 I learned what was known as QA – qualitative analysis. We would be given a mixture containing two or three chemical compounds and asked to identify the radicals[25] present. It was a complex, laborious, and rather smelly business that was already obsolescent at the time. I mention it because, in a limited sense, it harkened back to the days of alchemy, as we always started with the effects of water and fire: *Would the substances dissolve and what was the effect of heat?* If they dissolved, what were the effects of various reagents on the solutions? A great deal of the smell came from a thing called a Kipps Apparatus, which generated hydrogen sulphide and was often reluctant to stop when asked. H_2S, redolent of rotten eggs, was not part of the alchemists' arsenal, but fire and water were of the essence and remained so even as the older alchemical arts were lost and replaced by the beginnings of modern chemistry. Johann Baptista van Helmont (1579-1644), one of the pioneers of modern quantitative science, tells us that he dedicated himself to *pyrotechny*, by which he meant *chemistry*, and described himself as *philosophus per ignem* – a philosopher through fire. Some of his most important discoveries also involved working with aqueous solutions, but the tradition of the alchemical furnace persisted and Newton probably spent as many hours with his cauldron as he did with his quill. In the middle of the nineteenth century the Bunsen burner and the test tube became the outstanding popular symbols of the pursuit of chemical knowledge, but by then electricity had replaced fire as the most potent chemical tool, and electrical technology had become a force to be reckoned with.

In 1800, Alessandro Volta published reports of experiments he had done with plates of copper and zinc and strips of moist paper. He had, in fact, invented what became known later as the voltaic pile – the forerunner of all modern batteries. People had experimented with static electricity for thousands of years before this, but now scientists began to think in terms of electric currents – the "electrical fluid" moving along metal wires. In an amazingly short time the magnetic and chemical effects of these currents were investigated and put to use, and in 1827 Georg Ohm announced his famous law. In 1789, Lavoisier's list of chemical elements had included about thirty of the ninety or so that appeared in the periodic tables of the early twentieth century. Most of these had been isolated with the aid of fire and water, but now minerals that were stable at the highest available temperatures could be melted and torn apart by electrical currents. Using the new technique of electrolysis, Humphry Davy isolated six new elements[26] in the course of a few months. The reciprocal relationship between magnetism and electricity, investigated by Michael Faraday in England and by Joseph Henry in America, led to the construction of generators and of induction coils capable of producing very high voltages – in the thousands or even tens of thousands, as opposed to the hundreds available from batteries. These developments, together with improvements in vacuum technology, aided the continuation of a line of research that had begun more than a century earlier, namely the observation of what happens when high voltages are applied to highly rarefied gases.

Systematic investigation of these phenomena requires a reliable source of high-voltage electricity. Frictional machines for building up static electric charges were invented about 1700, some fifty years after the vacuum pump, and in 1709 Francis Hauksbee used both devices to demonstrate a precursor of the cathode ray tube and the neon light. William Watson seems to have been the first to produce the kind of experimental set-up that eventually became popular in the 1850's. In 1752 he was delighted to find that when he passed electricity through a partially exhausted tube, three feet long and three inches in diameter, the discharge passed much more easily than in air at atmospheric pressure and presented a very different appearance. The short, jagged sparks seen at normal pressure were replaced by flickering lights which ran the whole length of the tube.

Perhaps because of a long delay in the further development of the vacuum pump, no one seems to have followed up on Watson's remarkable discovery until Michael Faraday did so in 1838, but by the 1870's physicists all over Europe were working with vacuum tubes and induction coils and among the results were the discoveries of X-rays and the electron in the 1890's and the modified atomic theory of the early 1900's. That Rudolf Steiner was fully conversant with these developments is shown by his devoting a large part of Lecture 9 of the *First Scientific Lecture Course*[27] to them.

Meanwhile, Maxwell's theoretical demonstration that electrical and magnetic fields can interact to produce a wave motion the velocity of which is the same as the measured velocity of light, was confirmed in 1887 when Hertz produced radio waves in the laboratory, and more sensationally in 1895 when Marconi succeeded in transmitting radio signals over a distance of a mile.

The rapidly increasing importance of wireless telegraphy, as it was then called, is illustrated by an incident in the English Channel in 1899, in which the East Goodwin lightship was run down by a steamer and was able to communicate promptly with the South Foreland lighthouse, twelve miles away. This enabled lifeboats to reach the doomed lightship in time to save the crew. Two years later Marconi supervised the exchange of radio signals between Poldhu, in Cornwall, England, and St. John, Newfoundland. By the time of Steiner's last lectures on science, New Yorkers were adjusting their catswhiskers and tuning their crystal sets to a football game in Chicago, and Londoners were listening to the Savoy Orchestra on 2LO, a station shortly to be absorbed by the BBC. If this is not convincing enough of the pervasive presence of electricity in people's lives ninety years ago, we can add that by the mid-1920's, 13.5 million electric irons, 11 million vacuum cleaners and comparable numbers of toasters, washers and fans were already in use in the United States.[28] The situation was similar in Europe, and it wasn't only people's homes that were wired; many cities were lit by electricity and electric tramcars carried people about their daily business.

*

The discoveries of X-rays (Röntgen, 1895) and radioactivity (Becqerel, 1896), together with J. J. Thomson's demonstration in 1897 that cathode rays are streams of negatively charged particles, all contributed to a protracted debate about the atom, which now seemed certainly to be composite and not necessarily indivisible. These negative particles, which Thomson called corpuscles but which soon became known as electrons,[29] were much lighter than hydrogen atoms and were soon identified with the beta particles emitted by radioactive sources such as uranium. These sources also emitted much heavier particles – alpha particles – with positive charges, so the physicists found themselves facing the problem of assembling a model of the atom that incorporated something heavy and positive and something very light and negative. The shock of discovering that the atom was not, after all, the smallest particle of matter is evident in a comment made by Thomson in the course of a lecture to The Royal Institution in 1897: "The assumption of a state of matter more finely divided than the atom of an element is a somewhat startling one."[30]

(xi)

The Electrical Atom and Human Thought

For fifteen years after the discovery (or invention) of the electron, no one had any really convincing ideas about the structure of the atom. Lord Kelvin's model of 1902, consisting of a positively charged sphere with negatively charged particles – electrons – distributed through it like currants in a bun, was, for lack of anything better, fairly widely accepted. Punsters used to refer to the electrons as "electric currants." Steiner was almost certainly aware of this model in December 1904, when, in the course of a lecture[31] to the Theosophical Society, he referred to the atom as "coagulated electricity," explained the fundamental processes of evolution and involution, and linked them to the concept of the atom. In so doing, he brought out the critical need for selflessness. The difference between the tone of this lecture and that of the Goethe prefaces is connected with the fact that the latter were written for the general reading public, whereas the lecture was delivered in a smaller, more intimate, setting to an audience already familiar with the spiritual science Steiner was then in the early days of developing and articulating. Only one set of shorthand notes of the lecture survived; when the transcript was printed after the foundation of the Anthroposophical Society it was marked "Members Only."

God once gave us the nature that surrounds us in the kingdoms of the minerals, plants and animals. We take nature into ourselves. That nature exists is none of our doing; all we can do is to make nature part of our own being. But what we ourselves prepare and make ready in the world — that is what will constitute our future existence.

We actually see the mineral world, as such; what we do with the mineral world, that we shall ourselves become in future times. What we do with the plant world, with the animal world and with men, that too we shall surely become. If you found a charitable institution or have contributed something to its foundation, what you have contributed will become an integral part of you. If a man does nothing with what he can in this way draw into his soul from outside, then his soul remains empty. It must therefore be possible for mankind to spiritualize — as far as this can be achieved — the four kingdoms of nature, of which man is one....

We are living now in the epoch of evolution that may be called the mineral epoch; and our task is to permeate this mineral world through and through with the spirit within us. Think of what this means. — You are building a house. You fetch the stones from a quarry and hew them into the shapes required by the building, and so on. What are you inculcating into this raw material obtained from the mineral kingdom? You are inculcating human spirit into the raw material. If you construct a machine, you have laid the spirit that is part of you, into that machine; the actual machine does, of course, perish and become dust; not a trace of it will survive. But what you have done, what you have achieved, passes into the very atoms and does not vanish without a trace. Every atom bears a trace of your spirit and will carry this trace with it. Whether an atom has at some time been in a machine, or has not been in a machine, is not a matter

> *of indifference. The atom itself has undergone change as a result of having once been in a machine, and this change that you have wrought in the atom will never again be lost to it. Moreover, through your having changed the atom, through the fact that you have united the spirit in you with the mineral world, a permanent stamp has been made upon the general consciousness of mankind; just so much consciousness goes with you into the other world. Occult science well knows in what way the human being can perform selfless actions and how greatly his consciousness will be enhanced by them. Certain men, who have been deeply imbued with this knowledge, have been so selfless that they have taken steps to prevent even their names from going down to posterity... For selfless deeds are the real foundations of immortality...*

Thus far it might conceivably be argued that Steiner is using the word "atom" metaphorically, as when we speak of "every atom of a person's being" or say that a mirror is "shattered into atoms." However, the ideas of selflessness, the atom, and thinking come together again later in the lecture in a way that makes such an interpretation impossible:

> *All progress is the result of involution and evolution. Involution is the in-taking, evolution the yield, the out-giving. All states and conditions of world-existence alternate between these two processes. When you see, hear, smell or taste, you breathe nature into yourselves. The act of sight does not pass away without leaving a trace behind. The eye itself perishes, the object seen — that too perishes; but what you have experienced in the act of sight, remains. It will not be difficult for you to realize that in certain epochs it is necessary to make such things understood. We are going forward to an age when men will understand what the atom is, in reality. It will be realized — by the public mind too — that the atom is nothing but coagulated electricity. —*

Thought itself is composed of the same substance. Before the end of the fifth epoch of culture, science will have reached the stage where man will be able to penetrate into the atom itself. When the similarity of substance between the thought and the atom is once comprehended, the way to get hold of the forces contained in the atom will soon be discovered and then nothing will be inaccessible to certain methods of working.

— A man standing here, let us say, will be able by pressing a button concealed in his pocket, to explode some object at a great distance — say in Hamburg! Just as by setting up a wave-movement here and causing it to take a particular form at some other place, wireless telegraphy is possible, so what I have just indicated will be within man's power when the occult truth that thought and atom consist of the same substance is put into practical application.

It is impossible to conceive what might happen in such circumstances if mankind has not, by then, reached selflessness. The attainment of selflessness alone will enable humanity to be kept from the brink of destruction. The downfall of our present epoch will be caused by lack of morality. The Lemurian epoch was destroyed by fire, the Atlantean by water; our epoch and its civilisation will be destroyed by the War of All against All, by evil. Human beings will destroy each other in mutual strife. And the terrible thing — more desperately tragic than other catastrophes — will be that the blame will lie with human beings themselves.

In this present year of 2012, the attainment of selflessness, which seems as distant a goal as ever, may be far more important than the effort to come to a definitive understanding of Steiner's view of the atom, but this is not just an academic exercise. "The occult truth that thought and atom consist of the same substance," that thought and the atom are both electrical, is not something that we can just make a mental (or electrical) note of, before passing on to the next thing. We may agree that the nervous system, including the brain, is a hugely complicated electrochemical system, but this is not the same thing as saying that thought is composed of electricity. "Men will understand what the atom is, in reality. It will be realized — by the public mind too — that the atom is nothing but coagulated electricity and that thought itself is composed of the same substance." The words "understand" and "realize" indicate that what has been realized is something true, and Steiner refers to this as an "occult truth." Is there any way of understanding this?

As a tentative first step, we can recall that our thinking came originally from the spiritual world where it had existed before human beings took their present form. When the divine intelligence descended into the human soul and was gradually integrated into the human organism, the thoughts themselves were still potentially free but their expression came to depend on the functioning of the human physical body. As the centuries have gone by, our thinking has become more and more deeply embedded in the physical, but we have not altogether lost the potential to free it. That is one of the great goals of anthroposophical work. Asking what a thing is is not the same as asking what it is made of; Steiner does not say that thought is electricity any more than we would say that the Goetheanum is so many tons of concrete and colored glass. The matter is explained precisely by a "retired star" called Ramandu, in C. S. Lewis's *The Voyage of the Dawn Treader*:

> "In our world," said Eustace, "a star is a huge ball of flaming gas."
>
> "Even in your world, my son, that is not what a star is, but only what it is made of."

The Atom - A Historical Background — 49

*

In the period from 1919 to 1923 Steiner gave several scientific lecture courses, and in 1923 he spoke again about the electrical atom and its influence on human life. By this time Max Planck had found it necessary to invent the quantum, Albert Einstein had extended its applications, and Niels Bohr had designed the quantized electronic atomic model that became the mainstay of physics and chemistry textbooks for generations to come. By the mid-1920's, thanks to the work of Heisenberg, Born, and Schrödinger, among others, the quantum theory had begun to look like the basis for a theory of everything.

It is therefore tempting to write that if you want to understand what was happening in physics during the last few years of Steiner's life and after his death, you have to understand the quantum; the problem is that 111 years after its invention, there is still some doubt as to whether anyone understands it. It is possible, however, to get a good idea of where it came from, why it was necessary, what kind of a thing it is, and why it is important. If you are to appreciate Steiner's comments on atomic science and human consciousness near the end of his life, you need some of this background.

Chapter II
A Background for Quanta

(i)
Origins

The earliest form of quantum theory came out of a study of the energy radiated by hot objects and involved, among other things, a branch of physics known as thermodynamics. Thermodynamics, as practiced in the nineteenth century in the form known as "classical thermodynamics," dealt solely with the flow of heat and its relationship to mechanical work, and did not involve itself with atoms or molecules. This may seem paradoxical, since the notion of heat as the energy of motion of atoms and molecules had appeared quite early in the century. However, the physicist measures quantities of heat in terms that make no mention of any theory of matter – in other words, without making any attempt to say what heat is. One calorie, for example, is the quantity of heat needed to raise the temperature of a gram of water by 1 degree Celsius. Classical thermodynamics was therefore quite independent of the victories and defeats of the atomic theory and would not have been changed in any way if someone had suddenly discovered that the kinetic theory of matter was totally wrong. There was something god-like or, at least, oracular about its pronouncements – or so it seemed to me when I was a student in the 1950's.

Heat, as everyone knows, is a form of energy; but what is energy? As a humble student of spiritual science I can say, with some conviction, that it is one of the physical manifestations of the work of the hierarchies, that it comes from the spirit and that is why things sometimes happen that seem to fall outside our normal scientific expectations. Speaking as a physicist, I wish I could tell you what energy is, but if Richard Feynman couldn't, you can't expect me to be able to: "It is important to realize that in physics today, we have no knowledge of what energy *is*."[32] We have an idea about what it does; it enables trains, planes and automobiles to travel, rockets to go to the moon and falling bodies to reach lethal speeds. It is also associated with electrical currents that flow along wires, producing heat, and lead bullets that melt when they hit impenetrable barriers. We can calculate it for moving objects, electrical circuits, hot coffee and iced tea, and we believe that when it changes from one form to another none is created or lost, but we have no general conceptual model for energy. Nevertheless, as the nineteenth century wore on, the theory of energy developed a central role in the physical sciences, and people who had an inbuilt dislike for atoms tried to show that the use of energy theory made the atomic theory unnecessary.

The first and most basic idea of thermodynamics is that energy cannot be created from nothing and cannot be annihilated, so in any closed system there is a certain invariable amount of energy. If energy seems to appear or disappear, this is because it changes from one form to another. It is arguable that this generalization is simply a consequence of the way we measure work and energy and that it is not amenable to experimental proof. In practice, of course, there is no such thing as a perfectly closed system. We can try to imagine some gas in a container made of a perfectly insulating material, but there's no such thing as a perfect insulator, and even if we were able to construct such a vessel it would be impossible to make anything happen inside it and impossible to detect it if it did. So we are more apt to think of something like a bicycle pump with the exit hole blocked – in other words, a closed cylinder with a piston. When we push the piston in we do mechanical work and whatever is inside the cylinder gets warmer. The First Law of Thermodynamics simply says that there is a simple, constant relationship between the amount of work done pushing in the piston and the amount of heat produced. It doesn't matter whether the stuff inside the cylinder is air, sponge rubber or rusty nails – you do a certain amount of work and a certain amount of heat is produced. If you actually did an experiment like this, you would have to allow for heat escaping through the walls of the cylinder, but, like the Delphic Oracle, Classical Thermodynamics refuses to be involved with mere practical matters.

By the 1820's, pistons and cylinders had become very important. Newcomen's atmospheric engine was a century old and James Watt's reciprocating steam engine had passed its fiftieth anniversary. Stephenson's locomotives were already hauling coal, and in 1829 his Rocket began hauling people. It was clearly time for the physicists to get in on the act, and the French physicist Sadi Carnot did so in 1824 with his theory of the heat engine.

Carnot's engine is an idealized version of the cylinder and piston mentioned earlier. The bits and pieces are made of perfect conductors or perfect insulators so that they don't add any complications to the process; inside the cylinder is an elastic substance that we usually think of as a perfect gas. The "mental experiment" consists of four movements of the piston during which the walls of the cylinder are alternately perfectly conducting and perfectly insulating. This cycle of four strokes, not to be confused with the four strokes of a gasoline engine, became known as the Carnot Cycle, and with this bizarre and impossible concoction, he managed to prove something that actually applies to real steam engines and other heat engines, namely that there is a calculable upper limit on their efficiency no matter how perfectly constructed they are. This has to do with the fact, putting it very loosely, that energy is necessarily lost through the exhaust system unless this happens to be at the absolute zero of temperature.

What makes this early triumph of thermodynamics relevant to our study is that sixty years later the Carnot Cycle became an essential element in the theoretical work that eventually joined with the phenomenology of thermal radiation to make the quantum theory necessary.

(ii)

Thermal Radiation

As every student of general science learns, there are three ways in which hot objects lose heat – conduction, convection and radiation. Loss by conduction requires that the object be in contact with some cooler substance into which the heat can flow. If this cooler substance happens to be air, convection will take place too – warmer air will rise, taking heat with it. Meanwhile, the hot object is continuously radiating energy through the air. Removing the air around the object with a vacuum pump allows the radiation to proceed more efficiently. The observation that radiated heat passes through a vacuum at the same speed as light led to the conclusion that the two have the same form of propagation, namely electromagnetic waves of the kind described by Maxwell.

The most thermally radiant object in most people's experience is the sun, and if we pass a beam of sunlight through a prism, we find something else besides the familiar spectrum of light. A thermometer with a blackened bulb will show that radiant heat is present in the dark region next to the red, but we may need something a little more sophisticated to detect the ultra-violet radiation beyond the blue end of the spectrum. The term "thermal radiation" comprises this whole continuous, extended spectrum, including infra-red, visible light,[33] and ultra-violet. Note that what comes out of a fluorescent light bulb is not thermal radiation.

One observation vital for human health and safety is that, like hot terrestrial objects, the sun sends out far more energy in the infra-red than it does in the ultra-violet, and this brings us to the usual story that is told about the invention of the quantum. When Lord Rayleigh[34] applied the most widely accepted theories of the nineteenth century to the phenomena of thermal radiation, he found that the amount of energy emitted ought to increase without limit as we go further and further into the ultra-violet. Fortunately this doesn't happen – actually the amount of energy drops rapidly to zero. Rayleigh's formula was published in 1900 and, according to popular myth, Max Planck floundered around for a while before eventually putting the formula right by inventing the quantum theory. There are two problems with this story; one is that although it is historically simple, it involves some very complex physics, and the other is that it isn't true.

The great advantage of telling the true story in full would be that it shows that, like other theories that appeared in the nineteenth and twentieth centuries, the quantum theory was a response to otherwise intractable data, not a capricious or arbitrary invention. Sometimes, "under the compulsion of observation," to use Max Born's memorable phrase, physicists are forced to acknowledge inconvenient truths and to entertain ideas that they would prefer to know nothing about. Planck tried everything he could possibly think of and nearly drove himself crazy in the process before finally acknowledging that the quantum idea was the only one that worked. The disadvantage is that the story is so long and complex that hardly anyone would read it. I shall try to give the shortest intelligible version of the true story here, but be warned that "short" and "intelligible" are very elastic terms.

*

In what follows I shall use the terms frequency and wavelength quite often. What the physicist calls a complete wave consists of a trough and a crest.

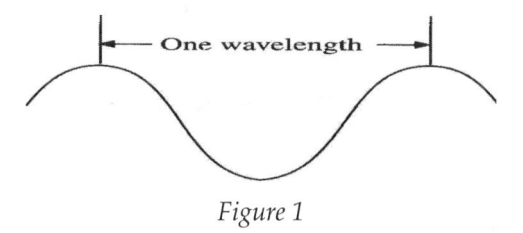

Figure 1

This applies equally to sound waves, which travel as alternating regions of high and low pressure, and electromagnetic waves, in which it is the field strength that alternates.

The *wavelength* is the distance between two consecutive crests, which is the same as the distance occupied by a trough and a crest. The *frequency* is the number of complete waves passing a given point in one second. The velocity of the wave is the product of the frequency and the wavelength, just as the velocity of a train is the product of the length of one car and the number of cars passing a given point in one second. For example, if each car is 50 ft. long and three cars pass in one second, the speed will be 150 ft/sec.

*

When Lord Rayleigh published his formula, physicists had already been studying the nature of radiant energy for many years, both experimentally and theoretically. The chain of observation and reasoning that led to the quantum theory stretches indefinitely into the past, but on the practical, observational side it is reasonable to start in 1859 when John Tyndall[35] began a series of experiments on the energy radiated by an electrically heated platinum wire. His main objective was to study the degree to which the radiant heat is absorbed by the different gases of the air, and in so doing he discovered the "greenhouse effect." He showed that the main atmospheric gases, nitrogen and oxygen, are almost transparent to radiant heat, whereas water vapor, carbon dioxide and ozone are such good absorbers that, even in small quantities, they absorb heat radiation much more strongly than the rest of the atmosphere.

These findings may turn out to be more important for the history of the world than all the theories of physicists put together, but it is a by-product of Tyndall's research that concerns us here. He published his results in 1865, and a few years later his book came to the attention of the great Slovenian physicist Josef Stefan (1835-1893). Stefan discovered that implicit in Tyndall's data was a connection between the temperature of a body and the rate at which it radiates energy: the rate at which energy is radiated is proportional to the fourth power of the absolute temperature. This obviously needs some explanation.

Data from several different kinds of experiment led the physicists of the nineteenth century to believe that there is an absolute zero of temperature. Measurements of the volumes, pressures, and temperatures of gases, taken at whatever temperatures the physicists could reach, indicated that if the arithmetical relationships of these quantities remained the same at lower and lower temperatures, the volume and the pressure would become zero at about −273°C. Since it was impossible to imagine a negative volume or a negative pressure, it was concluded that −273°C is the lowest possible temperature, or absolute zero. This was the basis of the Absolute Scale of Temperature, which took its zero as the absolute zero and measured all other temperatures from this point, while keeping the actual size of the degrees the same as in the Celsius scale. The melting point of ice is therefore 273°A and the boiling point 373°A.

In 1848, William Thomson, later Lord Kelvin (1824-1907, no relation to J. J. Thomson), created a theoretical (thermodynamic) foundation for the absolute scale, which subsequently became known as Kelvin or Thermodynamic Scale of Temperature. Absolute temperatures were then given in °K. At some point in the late twentieth century, some physicists decided to drop the degree symbol, so the melting point of ice is now written 273K. This is silly, since 273K could mean anything from a distance in kilometers to the size of a computer file or the price of a yacht; so I'll keep the degree symbol for the rest of this book. One of the great advantages of the Kelvin scale is that it makes the arithmetic of a great variety of physical relationships much easier.

Kelvin was most famous in his lifetime as the man who made the transatlantic cable work after two unsuccessful attempts by others. Partly as a result of this he became very wealthy and owned a private yacht with a fully equipped research laboratory on board. After retiring from his post as Chair of Natural Philosophy at the University of Glasgow at the age of 75, he showed his continued enthusiasm by immediately re-enrolling as a student.

*

The process of taking an observed relationship and continuing to apply it in regions that have not been explored is called extrapolation. Some of my friends have been known to speak derisively of the scientists' habit of extrapolation and to demonstrate its foolishness by giving silly examples, so it is as well to mention that the scientists have better reason to understand the perils of extrapolation than most of those who criticize them. I have heard it pointed out that just because Johnny grew six inches in his first year of life it doesn't mean that he'll grow fifteen feet in his first thirty years. Since these remarks often refer to extrapolation into the past, as in calculations of the age of the earth, it would be just as much to the point to say that if we have observed over the course of a comparatively short time that Johnny's height is staying the same, we have no reason to suppose that he wasn't always five feet seven inches tall.

After being regaled with such inanities we are invited to draw some conclusion, but I've never been quite sure what it is. It can't be that all extrapolations are wrong, since some have already turned out to be right. If the conclusion is merely that extrapolation is apt to give the wrong answer, all we can say is that we knew that already. But as exoteric scientists, if we want to find out about the distant past or future, or the behavior of matter under conditions that we can't reproduce in the lab, extrapolation is all we have. As Rudolf Steiner remarked, we must be prepared for the sun not to rise tomorrow morning, and as he also said:

> "As long as geology invented fabulous catastrophes to account for the present state of the earth, it groped in darkness. *It was only when it began to study the processes at present at work on the earth, and from these to argue back to the past, that it gained a firm foundation.*"[36]

In our daily lives we extrapolate all the time – otherwise we would never know what to do next.

*

As far as the arithmetic of Stefan's Law is concerned, a simple example will suffice. Think of a piece of platinum at 0°C; its absolute temperature is 273°K and it radiates heat at a certain rate. It is now heated to 273°C, so its absolute temperature is (273 + 273)° or 546°K. Since its absolute temperature has been multiplied by 2, its rate of radiation will be multiplied by 2 to the power of 4, or 16.

Using his fourth power law and measurements of the energy radiated by the sun, Stefan found that he could calculate the sun's surface temperature, which came out to about 5,500°C. Later workers with more accurate data put the figure at 6,000°C, and since the sun at high noon is the source of truly white light, this must be the temperature of a truly white-hot body.[37] Historically, the important thing to note is that no self-respecting physicist can learn of a new experimentally discovered law without wanting to know the reason for it.

*

Our next personality is Gustav Kirchhoff (1824-1887), who is well-known for formulating some of the fundamental laws of electrical circuits, for his work in spectroscopy, and for the discovery, with Robert Bunsen (of burner fame), of the alkali metals rubidium and cesium; but his most influential achievements may well have been those which concern us here: his discovery of the relationship between the power to emit radiation and the ability to absorb it, and his subsequent investigation of black-body radiation.

Substances vary in their ability to radiate energy. If, for example, you heat pieces of platinum, iron, and carbon each to 1,000°C and examine the radiation from them, you will find that they emit energy at different rates and that the energy is distributed differently over the different parts of the spectrum. Kirchhoff showed that the ability to emit radiation is closely related to the ability to absorb it. Sodium vapor, for instance, emits a strong yellow light when agitated, and when cool and undisturbed it is a very good absorber of the same color; to put it very simply, good radiators are also good absorbers. A perfectly black body would absorb all the radiant energy that fell on it and would, according to this generalization, be a perfect emitter of radiation. Kirchoff then sought to create a perfectly black body, approaching the problem by making a kind of oven – a box with the interior coated with graphite, which is the most nearly perfect absorber and emitter of radiation. If the walls of the box are maintained at a uniform temperature, the character of the radiation inside depends only on the temperature. As a result of further experiments and some heavy thinking, he reached the remarkable conclusion that it doesn't matter what the box and any objects inside it are made of.

A Background for Quanta — 61

Such a box of radiation is known as a uniform temperature enclosure. All we have to do is to make a little hole in the wall of the box and maintain the walls at whatever temperature we desire and we have a source of what we might call perfect radiation. (Reminder; we are talking about radiant energy, usually thought of as heat and light. This has nothing to do with radioactivity.) Now it was possible to investigate the radiation, both theoretically and experimentally.

One of the great theoreticians who continued the work on thermal radiation was Ludwig Boltzmann (1844-1906), who is worth studying not only for his contributions to theoretical physics.

*

Boltzmann was born in Vienna in 1844, studied at the University of Vienna, where Josef Stefan was one of his tutors, and succeeded Stefan as professor of theoretical physics in 1894. After a short spell in Leipzig, he returned to Vienna to resume his old chair and also the chair of natural philosophy recently vacated by Ernst Mach. All of his work, which included vital contributions to thermodynamics, was based on a firm belief in the physical existence of atoms and molecules. Although there was increasing acceptance of this in much of the scientific world, at the end of the nineteenth century there was still strong opposition in Vienna, particularly from Mach. Boltzmann's life and death, and his interaction with Mach, give an important insight into the debate over the atomic theory that had been going on for much of the nineteenth century.

In 1840, when Mach was two years old, his family moved from Moravia, which was then part of the Austro-Hungarian Empire, to a small town near Vienna. After receiving his Ph. D. from the University, he spent twenty-eight years as professor of experimental physics at the University of Prague, eventually returning to Vienna in 1895 as professor of history and philosophy of science. He was a profound thinker, some of whose ideas made a deep impression on the physicists of his time; what interests us here is his opposition to the atomic theory, which was based on the principle that anything that is imperceptible to the senses cannot be regarded as real. This sounds highly reminiscent of Goethean science, so we must realize that although Mach was in some ways a maverick, his physics was still a product of the nineteenth century scientific ethos. He wasn't what we might call a gentle scientist.

Mach and Boltzmann were therefore in conflict over the matter of atoms and, the situation in Vienna being what it was, Boltzmann felt that he was fighting for a lost cause. Suffering from depression and unaware that Einstein's work on Brownian motion had convinced many of the remaining sceptics of the truth of the atomic theory, he committed suicide in 1906.

Most people were not as passionate, committed, emotionally involved and psychologically troubled as Boltzmann, but the debate certainly aroused very strong feelings on both sides and, like many other scientific controversies, was not always conducted in the kind of clinical atmosphere that people sometimes associate with the interactions of physicists.

*

One of the conclusions from Maxwell's work is that if radiant energy is indeed carried by electromagnetic waves, it must exert a pressure somewhat analogous to the pressure of a gas.

A Background for Quanta — 63

This being the case, Boltzmann had the idea that something useful might follow if he imagined one of Kirchhoff's radiation boxes transformed into the cylinder and piston of a heat engine like the one designed by Carnot. In Boltzmann's mental experiment the radiation expands and contracts as the piston moves back and forth. Applying Carnot's result for the efficiency of an ideal heat engine, and mathematics not beyond many present-day high school students, Boltzmann was able to provide a theoretical derivation of Stefan's Fourth Power Law of Radiation. Bear in mind that according to Boltzmann's ideas, this is a property of the radiation itself, not of any material substance, and doesn't depend on any theory about how the radiation is produced.

So Boltzmann had done for Stefan's Law what Newton and Bernoulli had done for Boyle's Law in the seventeenth and eighteenth centuries. They had assumed a set of properties for the atoms it was supposedly composed of and calculated the result; Boltzmann assumed that the radiation had the properties described by Maxwell and showed that the Fourth-Power Law was the result.

This was, you may say, satisfactory, but it only dealt with the energy as a whole, not with the way it was distributed over the different parts of the spectrum – how much in the ultraviolet, the visible spectrum and the infra-red. There was plenty of experimental information about this distribution – in general the energy falls to zero in the deep infra-red and the far ultra-violet and peaks somewhere in between. The following diagram shows graphs of the energy distribution at temperatures of 3,000, 4,000, 5,000 and 6,000°K. As the temperature rises, the peak moves further and further to the left – that is, towards the ultra-violet – and the peaks, labeled λ_{max}, lie on a smooth curve shown by a dotted line.

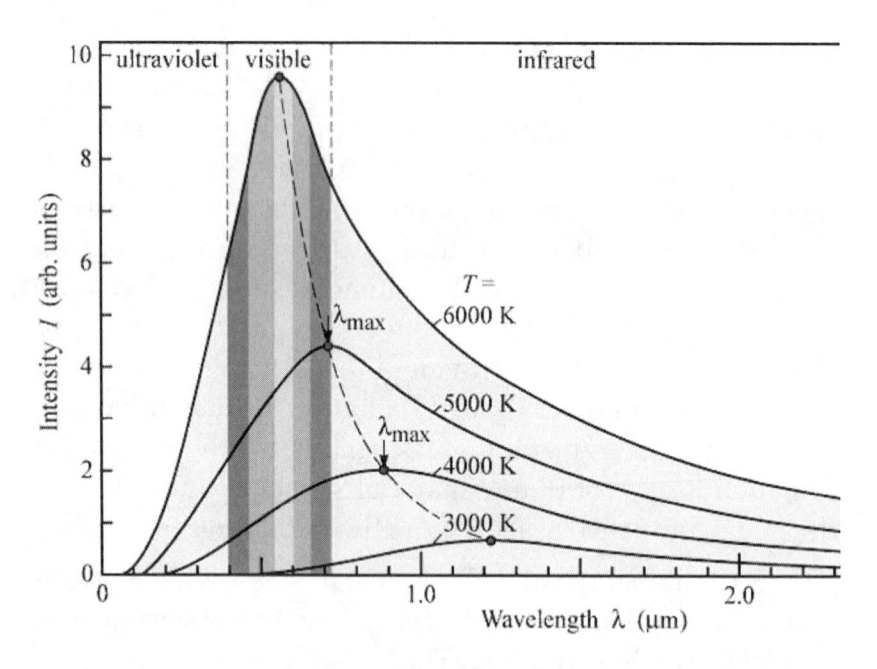

Figure 2:
Distribution of energy over the spectrum of blackbody radiation

Now along came Wilhelm Carl Werner Otto Fritz Franz ("Willy") Wien (1864-1928), who was one of the greatest and least appreciated participants in the long struggle to understand the physical nature of radiant energy. Wien realized that the aforementioned smooth curve (the dotted line on the diagram) is a hyperbola governed by a very simple rule – Wien's Displacement Law, as it became known: when you multiply the peak energy by the wavelength at which it occurs, you always get the same answer, whatever the temperature. Using a method similar to Boltzmann's, he was able to go further and deduce this rule theoretically.

This was a big step forward, but the physicists would not be satisfied until they had found not only a formula that governed the shape of the whole energy distribution but also a theoretical explanation of it. You may wonder why this seemed so important that people were willing to spend major parts of their lives on it without any assurance that their quest would be successful. There is no quick answer to this question, but the beginnings of a response may be found in the epilogue to Sir Arthur Eddington's *New Pathways in Science*;[38] "Whatever else may be in our nature, responsibility towards the truth is one of its attributes." Some of us may object to the ways in which others pursue the truth, but the pursuit is, in itself, admirable.

(iii)

Enter Max Planck

The successes that I have described were achieved by considering the nature of the radiation itself without making any assumptions about the mechanisms by which it is generated, but the situation changed when Wien tried to apply some new methods developed by Maxwell and Boltzmann in their work on the kinetic theory of matter. Earlier workers in this field had achieved their results by adding up the effects of all the individual particles but Maxwell and Boltzmann used probability theory to predict the behavior of large assemblies of atoms or molecules; in other words, they treated the particles statistically instead of individually. As Wien reported, "I tried to get around the problem of making probability calculations about radiant energy by imagining the radiation as resulting from gas molecules moving according to the laws of probability..."[39]

Wien was, in fact, remarkably successful. In 1896, a year before the discovery of the electron, and four years before Lord Rayleigh's formula appeared, he produced a formula that seemed accurately to describe the whole family of curves in *Figure 2* and whose only deficiency was that it lacked a convincing theoretical background. This was the situation when Max Planck began his greatest battle, and several more years elapsed before experimental physicists found that Wien's formula, although accurate in the ultra-violet and visible regions of the spectrum, became seriously inaccurate in the deep infra-red. This is the part of the story that my teachers omitted, perhaps because they weren't aware of it or perhaps because it didn't fit in with their ideas of the logic of the situation. We were taught that Planck introduced the notion of the quantum in order to correct a problem in the ultra-violet (the very short wavelengths on the left side of the graph), as revealed in Rayleigh's formula. In fact, Wien's formula worked admirably for the ultra-violet and for the visible spectrum, but not so well for the infra-red.

To sum the situation up: Planck was at first under the impression that Wien's formula gave an accurate representation of the experimental facts and his main concern was to provide a sound theoretical basis for it as part of a whole theory of radiation with some unifying and predictive power. The revelation that the formula was inaccurate in the infra-red added to Planck's difficulties. Rayleigh's formula of 1900 might have been a help or a hindrance, but it was not part of Planck's thinking when he introduced the quantum.

*

Max Planck (1858-1947) was born in Kiel, a few miles south of the border between Prussia and Denmark. As a child he saw Prussian and Austrian soldiers marching into his home town during the Danish-Prussian war; an omen, perhaps, of the tragic events of his later life. Planck was a very musical youth, singing, playing several instruments and composing songs and operas, but after graduating from the Gymnasium in Munich, where his family had moved in 1867, he chose physics as his career. He studied the work of Clausius, one of the great authorities on kinetic theory and thermodynamics, and became deeply interested in the theory of heat. After a slow start, his academic career took off and in 1892 he was appointed professor of theoretical physics at the University of Berlin. By this time he was married and had three children, a fourth arriving a year later. Planck's friends and colleagues, including Albert Einstein, were always welcome in his home and, as is often the case among scientists and their friends, there was a lot of music-making.

The family's happiness did not last; Planck's wife died in 1911, his oldest son was killed in action in 1914, and his second son, Erwin, was captured by the French. Both his daughters died in childbirth, and Erwin survived the First World War only to be executed by the Nazis in 1945 after taking part in a plot to assassinate Hitler. Stoically enduring all these losses, Planck lived to a great age, honored and revered by scientists all over the world. It is noteworthy that Max Born placed him firmly in the line of the ancient philosophers who believed in the primacy of thinking; "He was by nature and by the tradition of his family conservative, averse to revolutionary novelties and sceptical towards speculations. *But his belief in the imperative power of logical thinking based on facts was so strong that he did not hesitate to express a claim contradicting all tradition*, because he had become convinced that no other resort was possible."

*

The situation encountered by Planck was well described by both Wien and Planck in their respective Nobel Prize Lectures.

Wien: "As soon as we step beyond the established boundaries of pure thermodynamic theory, we enter a trackless region confronting us with obstacles which even the most astute of us are almost at a loss to tackle..."[40]

Planck: "When I look back to the time, already twenty years ago, when the concept of the physical quantum of action[41] began to unfold from the mass of experimental facts, and again, to the long and ever tortuous path which led, finally, to its disclosure, the whole development seems to me to provide a fresh illustration of the long-since proved saying of Goethe's that man errs as long as he strives. And the whole strenuous intellectual work of an industrious research worker would appear, after all, to be vain and hopeless, if he were not occasionally through some striking facts to find that he had, at the end of all his crisscross journeys, at last accomplished at least one step which was conclusively nearer the truth."[42]

Planck's lecture shows his generosity and modesty. Those who had gone before him in the search for understanding are warmly acknowledged, and he shows his gratitude for the help of Ludwig Boltzmann, who corrected him at one point when he had set off on a false trail, and whose letter of agreement at a later stage brought him some much-valued satisfaction after many disappointments.

In trying to understand how Planck went about his work, we should bear in mind that he placed himself in a long line of scientists who calculated the behavior of theoretical models with idealized objects and substances and found that their conclusions had some applicability in the real world. Newton had imagined a fluid consisting of "...particles fleeing from each other, with forces that are inversely proportional to the distances of their centres..."[43] and although he plainly stated that he didn't know whether

A Background for Quanta — 69

air was really constituted in this way, his model predicted not only Boyle's Law but also the actual way in which the atmospheric pressure varies with the altitude. (Physicists, it should be noted, have a habit of using their theories to predict things that they already know. Somehow this reminds me of Hobbits, who "like to have books filled with things that they already know, set out fair and square with no contradictions.")[44] And Newton's calculation gave the same answer as a later one, equally a mental experiment, in which it was assumed that a gas consists of particles that simply ignore one another. Carnot's heat engine was a thought experiment that can be regarded as the grandfather of the sequence in which Boltzmann derived Stefan's Law, Wien derived his displacement law, and Planck struggled and finally succeeded where Wien had failed in creating a distribution law for thermal radiation. These methods don't necessarily prove anything about the inner nature of substance, but they do get results. Carnot's Theorem sets an actual limit on the efficiency of real heat engines. The radiation engines of Boltzmann and Wien gave exact predictions of two experimentally verifiable laws of radiant energy, and Wien obtained a promising approximation to curves in *Figure 2*. Eventually, however, it became clear that such visualizations have their limits.

In the early days of the kinetic theory it was easy to make a mental image of the molecules of a gas incessantly milling around "like motes in a sunbeam," to borrow a phrase from Lucretius, and bumping into each other and the walls of the containing vessel. Striking results could be obtained with mathematics that didn't go beyond the modern high school level. Carnot's engine, with its frictionless piston and its perfect conductors and insulators, is more obviously an impossible dream, but it is easy to visualize, generates no difficult mathematics and has clear applicability to actual situations. When we come to boxes with adjustable, perfectly reflecting or absorbing walls, full of radiant energy, visualization gets a little harder but not impossible; but the use of statistical methods places us at a further remove from the good old days when we could imagine atoms as individual objects interacting with one another. Planck's theoretical model, which I'll do my best to explain up to a point, introduced another obstacle to easy visualization and, as we shall see, by the mid 1920's the visualization of atomic and subatomic processes had become not only something between very difficult and impossible, but also potentially misleading.

*

In Wien's mental experiment, the waves of radiant energy came from the random emissions of gas molecules or electrons. Planck decided on something that is even harder to imagine. Since all the waves were believed to be electromagnetic, why not imagine that they came from tiny antennas radiating waves like the ones demonstrated by Heinrich Hertz in 1888? In fact, you can probably think of several reasons for not doing this! The simple mental image that many people now have of an atom with electrons circulating around a nucleus may make it easy to imagine atoms acting as radio antennas but in the 1890's there were no such models of atomic structure. There were, however, good reasons for Planck to make this idea the basis of a trial run.

> *If a number of such oscillators [tiny radio stations] are set up within a cavity surrounded by a sphere of reflecting walls... the so-called black-body radiation should be set up within the*

> *cavity. I was filled at that time with what would be thought today naively charming and agreeable expectations, that the laws of classical electrodynamics would... be sufficient to enable us to achieve the desired aim.*

The "desired aim" was to find a theory that would be in exact agreement with Wien's radiation formula, but what Planck obtained was "no more than a preparatory step towards the initial onslaught on the particular problem which now towered with all its fearsome height even steeper before me."

Planck's first assault on the "fearsome height" was based on some mistaken assumptions, one of which "was strongly contested by Ludwig Boltzmann... with his riper experience in these problems.... All these analyses showed ever more clearly that an important connecting element... must be missing."

Having tried and failed to dream up a mechanism that would give the right result, Planck decided that he'd better go back to good old thermodynamics:

> *So there was nothing left for me but to tackle the problem from the opposite side, that of thermodynamics, in which field I felt, moreover, more confident.*

His next remark, however, may make the reader feel much less confident.

> *In fact my earlier studies of the Second Law of Heat Theory stood me in good stead, so that from the start I tried to get a connection, not between the temperature but rather the entropy of the resonator and its energy...*

So what is entropy and where did it come from?

*

When God created the heavens and the earth (Genesis I), He performed several acts of separation. The earth was originally "without form and void," without any kind of order. A physicist might say that this was a condition of maximum entropy, since entropy is a measure of disorder. In separating the light from the dark, the upper waters from the lower waters and the dry land from the ocean, God accomplished an increase in order and therefore a decrease in entropy. A physicist might argue that the only possible explanation is that God's entropy must have increased, but this only emphasizes the gulf between the divine and the human.

Human operations always seem to be accompanied by an increase in entropy. We may think that by building a house we have brought more order into the world, but one look at the building site during the process of construction may well convince us otherwise. After the debris has been carted away and the house is surrounded by lawns and gardens, we breathe a sigh of relief and contemplate the order we have created. We probably don't calculate the volume of random landfill and the amount of mechanical and electrical energy that has been used, much of which has simply escaped into the environment as heat, producing a large increase in entropy. Energy that was available to drive a drill, a chain saw or a cement-mixer is now dispersed in the atmosphere; an increase in entropy doesn't decrease the amount of energy – it just spreads it out randomly and makes it unavailable for anything useful.

Entropy is hard to quantify in absolute terms, so physicists usually deal only with changes, calculated in terms of the quantity of energy involved and the temperature. As you will realize if you think about the story as so far related, all the investigators up to and including Planck were preoccupied with quantities of energy. Having already done an enormous amount of work on the problem, Plank realized that to make the final step of finding a theoretical basis for the observed spectrum of black body radiation, he had to shift his focus to entropy. Considering that changes in entropy correspond to changes in the degree of randomness in a system, it's not unreasonable to suggest that entropy is an appropriate concept when we are trying to solve problems with a statistical approach that deals in probabilities rather than cause and effect.

This explanation gives only a general idea of the possible relevance of entropy to the problem of thermal radiation. When we continue Planck's remark that I interrupted a moment ago, we realize that math anxiety is fully justified and that we passed the limit of non-technical explanation some time ago.

> ...and in fact, not its entropy exactly but the second derivative with respect to the energy since this has a direct physical meaning for the irreversibility of the energy exchange between resonator and radiation.[45]

However, we can still get some idea of how Planck's thinking went on. Not fully grasping the importance of the "connection between entropy and probability, I saw myself, at first, relying solely upon the existing experimental results. In the foreground of interest at that time, in 1899, was the energy distribution law established by W. Wien a little earlier..."

Wien's formula was, as you remember, correct for short wavelength radiation (ultra-violet) and increasingly inaccurate for long wavelengths (infra-red). Splitting the problem into two parts, Planck worked out a relationship between entropy and energy that worked for short wavelengths, and from the experimental results for long wavelengths he found a different relationship. He then found that he could get a formula that worked for all wavelengths by suitably combining the two relationships. It was nice to have a formula that was correct for the whole range of wavelengths but, of course, Planck wanted to know *why*. There must be a physical concept from which the formula could be generated rather than getting it from arbitrary-seeming combinations of experimental results and mathematical trial and error.

> *For this reason, I busied myself... with the task of elucidating a true physical character for the formula, and this problem led me automatically to a consideration of the connection between entropy and probability... until after some weeks of the most strenuous work of my life, light came into the darkness, and a new undreamed-of perspective opened up before me.*

Planck explains that his formula involved two numbers that can be described as universal constants – numbers that are always the same no matter under what conditions they are measured.[46] The second of these, which was later given the name *Planck's constant*, could be determined from the experimental results, but Planck, as always, wanted a physical reason for its magnitude. This was the cause of what he later described as "an act of desperation." Having struggled for six years with the problem of black-body radiation, he felt that a theoretical explanation had to be found at any cost, however high – short, that is, of abandoning the laws of thermodynamics. The cost was indeed high – the abandonment of one of the most obvious assumptions about electromagnetic waves, namely that they are emitted in continuous streams. The *"act of desperation" was the assumption that radiant energy is emitted discontinuously in separate packages ("certain quanta"), the size of which is proportional to the frequency of the radiation.* In broad terms this means that ultra-violet (shorter wavelength and higher frequency) comes out in relatively large packets and infra-red (longer wavelength and lower frequency) in relatively small ones. All the packets, however, are sub-microscopically small in comparison to everyday quantities of energy, like the calorie, their actual magnitude being the product of the frequency (cycles per second, now known as Hertz, Hz) and Planck's constant. Physicists traditionally used the Greek letter ν (nu) for frequency and Planck's constant was given the letter h, so the energy of one quantum is given by

$$E = h\nu$$

It's a pity that ν looks so much like v; in fact, the italic ν (*ν*) seems to be identical with the italic v (*v*), but to the physicist it's a matter of context. Anything that looks like a v (vee) must be a ν (nu) if it follows an h. Many modern texts, however, have given up on ν and write

$$E = hf$$

(Using f for frequency may seem to be a no-brainer – the difficulty was that in my youth there were still middle-aged and elderly mathematicians and physicists who used f for acceleration. One way around this is to use n instead.)

Planck didn't like his quantum very much and the only thing he used it for was, "under the compulsion of observation," to derive a distribution equation for thermal radiation that fitted the experimental facts. So it was an *ad hoc* assumption with no basis in existing theory, and no one at the time could foresee that it had a brilliant future. As

Planck pointed out in his Nobel Prize lecture, there was a serious question "whether the whole method is to be looked upon as a mere artifice for calculation, or whether it has an inherent real physical meaning and interpretation."

> *Either the quantum of action was a fictional quantity, then the whole deduction of the radiation law was in the main illusory and represented nothing more than an empty non-significant play on formulae; or the derivation of the radiation law was based on a sound physical conception. In this case the quantum of action[47] must play a fundamental role in physics, and here was something entirely new, never before heard of, which seemed called upon basically to revise all our physical thinking, built as this was... upon the acceptance of the continuity of all causative connections.*

> *Experiment has decided for the second alternative. That the decision could be made so soon and so definitely... should be attributed to the restless forward-thrusting work of those research workers who used the quantum of action to help them in their own investigations and experiments. The first impact in this field was made by Albert Einstein...*

*

Within five years of Planck's publication, Einstein had used the quantum to account for some otherwise inexplicable observations of the electricity – always referred to as electrons – released when light falls on the surface of certain metals. This is known as the photo-electric effect and people were soon using photo-electric cells as triggers for automatic doors. The difference between Einstein's use of quanta and Planck's was that whereas Planck had reluctantly used them as a device to account for the phenomena of the emission of radiant energy, Einstein began to feel that it was necessary to assume that the energy always existed as quanta, whether it was just setting out on its journey, traveling or arriving at its destination. Most of Einstein's colleagues in the scientific community acknowledged the accuracy of his mathematics; but even when he used his theory successfully to solve several other problems, including those of low temperature physics described by Maxwell in 1869, they still found that his version of quantum theory was far too radical to be accepted. Einstein himself regarded his explanations as provisional, since they did "not seem reconcilable with the experimentally verified consequences of the wave theory," and one of the most striking episodes in this tale is the experience of the great American experimental physicist Robert Millikan (1868-1953), who was also a great sceptic:

> *I spent ten years of my life testing that 1905 equation of Einstein's, and contrary to all my expectations I was compelled in 1915 to assert its unambiguous verification, in spite of its unreasonableness.*

As Helge Kragh[48] points out, "what Millikan had confirmed was Einstein's equation, not his theory," which Millikan still thought of in 1916 as "a bold, not to say reckless hypothesis."

*

I have already spoken of Max Planck's musical talents, outlined the tragic and heroic course of his life and emphasized his intellectual honesty; now I feel that it is only fitting to add something about his moral rectitude, since it is not unknown for a writer to ignore the evidence for his inner goodness and dwell on his so-called "unsavory compromises" with the Nazi regime.

Planck has often, and rightly, been described as a conservative. The quantum offended his sense of the rightness of classical physics,[49] and his conservatism extended to social matters, including the position of women in society. Voicing the opinions of male establishments all over the world, he wrote in 1897, "... it cannot be emphasized strongly enough that Nature has designated for woman her vocation as mother and housewife, and that under no circumstances can natural laws be ignored without grave damage..."

He was, however, quite capable of forgetting his social conservatism in order to respond to the needs of an actual situation. One of only two women deeply and visibly involved with the major developments in atomic physics in the first thirty years of the twentieth century was the great experimental physicist Lise Meitner (1878-1968) – the other, of course, being Marie Curie. Although Meitner had received a doctorate in physics from the University of Vienna in 1906, for the next six years she was unable to find a paying position as a research scientist. As Gino Segrè puts it in his indispensable *Faust in Copenhagen*, "Concerned about her, particularly after the death of her father, Planck did something extraordinary, a move Meitner always regarded as the turning point of her career. In 1912 he appointed her as his *Assistent*, her first paid position and the first woman *Assistent* in Prussia."[50] From that point her career took off, and, working with Otto Hahn, she pioneered the study of nuclear fission. It is typical of the position of women scientists in that era that Hahn was awarded a Nobel Prize for his contribution while Meitner was overlooked. Half a century later she wrote of Planck:

> *He had an unusually pure disposition and inner rectitude, which corresponded to his outer simplicity and lack of pretension... Again and again I saw with admiration that he never did or avoided doing something that might have*

been useful or damaging to himself. When he perceived something to be right, he carried it out without regard to his own person.

Planck was devoted to Germany and its institutions but, quoting Segrè again, "at the same time he continually questioned these loyalties and often struggled with their implications, certainly during the First World War and again as Nazism arose in his country. Meitner was correct in emphasizing how Planck always tried to do what he thought was right without regard for his own welfare."

*

There were many more surprising developments in the period before Steiner's next sequence of scientific lectures, including Niels Bohr's application of quantum theory to the structure of the atom, but I think that the reader is entitled to a period of rest from quanta while we ponder Steiner's thoughts about the science of his time.

Chapter III
Steiner in the Quantum Age

(i)

Physical Science and Spiritual Science

When I first encountered Rudolf Steiner fifty years ago I was given the impression that anthroposophists disapproved of almost everything that was modern. Twentieth century art and science seemed to be generally regarded as grossly and unrelentingly ugly and materialistic – the products of people entirely cut off from the spiritual world. It took some time for me to realize that these opinions were based largely on personal feelings and represented a very one-sided knowledge and interpretation of Steiner's view of the world. It is true that artists and scientists are infected with materialism, but so, of course, is everyone else, including anthroposophists.

People relying on English translations had some excuse for being ignorant of Steiner's scientific writings, but things began to change with the 1969 publication in English of Karl Stockmeyer's great work on the curriculum of the Waldorf Schools.[51] I was immediately struck by Steiner's insistence that an adequate preparation for life must include a knowledge of modern developments in physical science. At a meeting with the first Waldorf teachers in 1919 he specifically mentioned wireless telegraphy, X-rays and alpha, beta and gamma rays, which were newer to the scientists then than quarks and chaos theory are now, and spoke eloquently about the need for people to understand the technological workings of modern life.

> We live in a world produced by human beings, moulded by human thought, of which we make use and which we do not understand in the least. This lack of comprehension for human creation, or for the results of human thought, is of great significance for the entire complexion of the human soul and spirit...[52]

An important event for me was the publication in 1982 of *The Boundaries of Natural Science*, the translation of a series of lectures given in 1920, in which Steiner links the knowledge of modern scientific method with the development of spiritual science.

> ...nobody can attain a true knowledge of the spirit who has not acquired scientific discipline, who has not learned to investigate and think in the laboratories according to the modern scientific method. Those who pursue spiritual science have less cause to undervalue modern science than anyone. ...it is the quality of this scientific method and its results that we must take very seriously indeed.[53]

Although this is not to be taken as an unqualified endorsement of modern science, it is certainly a strong warning against ignorance and disdain. I had to wait another nine years for the publication of *Anthroposophy and Science*, a course given in 1921. Here Steiner deals with the mathematical treatment of the natural world, points to the feeling of certainty which arises from this, and remarks that the cause of this feeling is not immediately obvious. He continues:

> *A clear knowledge of the feeling that accompanies the use of mathematics [in natural science] will lead us to acknowledge the necessity that a spiritual science must come about with an equal degree of certainty....* This spiritual science will conform in every discipline to the scientific consciousness of the times; it will, in addition oppose all that is brought forward by modern science that is suspect, and it will answer questions that often go unanswered. Spiritual science will be on a very sure mathematical foundation.[54]

What a spiritual scientist wishes to build on that foundation is something that goes beyond the mundane certainties of science as it appeared in 1920, and does not carry with it the usual penalty – loss of the immediate experience of the natural world.

"Especially in our age," Steiner says, "in which there is real proficiency in the handling of facts in an outer experimental way... what is investigated experimentally *must be permeated with the results of spiritual research.*" Our experimental, mathematical way of dealing with our surroundings brings "order and harmony into the otherwise chaotic stream of everyday facts... [but] *one has to admit that all the knowledge obtained in this way stands as a closed door to the outer world in that it does not allow the essence of this outer world to enter our cognition...* As long as we remain in this field of knowledge, we cannot see through the outer appearances; we also, of course, do not claim to do so... Basically we need this kind of knowledge to maintain our normal human consciousness."

The maintenance of our normal human consciousness in a healthy state is a prerequisite of any kind of legitimate spiritual endeavor. The ladder has to have something firm to stand on. The next question is:

> How does one arrive at anthroposophical spiritual science when starting from the familiar science of the present day? I don't believe anyone will be able to answer this question in a truly scientific way who cannot first answer the question: How is our observational knowledge raised to the kind of knowledge that is permeated with mathematics?

The mathematical treatment of nature helps to give us a feeling of unity with an outer world that otherwise seems foreign. But the picture we have created no longer contains the reality which presented itself to us originally. The abundance of sense impressions – colors, sounds, smells, tastes, textures – is lost, and nothing in the whole world of mathematical representation can replace it. Mathematical knowledge takes us more deeply into our inner life, and we feel that it should help us to a more intimate experience of the outer world. But when we impose our reassuring inner mathematics on the outside world, we cut ourselves off from the world's essential being. The equations which enable us to design optical instruments and explain rainbows give no inkling that there is such a thing as the actual sensation of color.

The crucial questions which I have quoted above can now be put in a more forward-looking form:

> Is it possible that what is first experienced mathematically as pale abstractions can be made stronger? In other words, could the force which we have to use to attain a mathematical knowledge of nature be used more effectively, with the result not just a mathematical abstraction, but something inwardly, spiritually concrete? ...This we can see as a third step in attaining knowledge. The first step would be the familiar

> *grasping of the real outer world. The second would be the mathematical penetration of the outer world, after we have first learned inwardly to construct the purely mathematical aspect. The third would be the entirely inner experience, like the mathematical experience but with the character of spiritual reality.*

To take the third step is the object of every spiritual seeker, but the temptation to omit the second one becomes stronger as the years go by. Whatever smacks of the intellect, especially the scientifically oriented intellect, encounters the doubts, misconceptions and prejudices of those who distrust anything appearing to come from that source. Steiner said that he did not expect all anthroposophists to start hitting the math texts, but he did insist that clear thinking is where everything starts. "There is no denying that *before anything else can be understood, thinking must be understood.*"

*

The Origins of Natural Science was the third of the great public scientific lecture courses given by Steiner in the period 1920-1923. In the second lecture, given in Dornach on Christmas Eve, 1922, he says, "*…The scientific path taken by modern humanity [is] … not erroneous but entirely proper… It bears within itself the seed of a new perception and a new spiritual activity of will.*"[55]

He has, however, already warned us to proceed with caution. While it is true that scientific thinking cannot proceed at all unless it is propelled by will activity, the result of this propulsion is a momentum which carries the process beyond the phenomena and into the world of atomic constructs.

> *Out of the clear concepts I have achieved I shall construct atoms, molecules — all the movements of matter that are supposed*

to exist behind natural phenomena. Thereby something extraordinary happens. What happens is that when I as a human being confront the world of nature, I use my concepts not only to create for myself a conceptual order within the realm of the senses but also to break through the boundary of sense and construct behind it atoms and the like. I cannot bring my lucid thinking to a halt within the realm of the senses. I take my lesson from inert matter, which continues to roll on even when the propulsive force has ceased. I have a certain inertia, and I roll with my concepts on beyond the realm of the senses to construct there a world the existence of which I can begin to doubt when I notice that my thinking has only been borne along by inertia.

It is interesting to note that a great proportion of the philosophy[56] that does not remain within phenomena is actually nothing other than just such an inert rolling-on beyond what really exists within the world. One simply cannot come to a halt. One wants to think ever farther and farther beyond and construct atoms and molecules—under certain circumstances other things as well that philosophers have assembled there. No wonder, then, that this web one has woven in a world created by the inertia of thinking must eventually unravel itself again.[57]

By 1920, the time of these last remarks, the web of atoms and molecules, as understood by the pre- and early-quantum physicists, was already becoming unraveled, and soon quantum physics would be the path for all mainstream fundamental research to follow, leading into a strange and confusing world where most of the signposts of old-fashioned physics were missing. There was, however, a path less traveled that had led Steiner, years earlier, to a Goethean science in which precise thinking dwells on phenomena and seeks necessary connections through contemplation rather than manipulation.

(ii)
The Goethean Alternative

In 1918 Steiner wrote a new forward for the second edition of *Goethe's World View*,[58] originally published in 1897, stressing that there had been no significant change in his understanding of Goethean epistemology over the preceding twenty years. The reason for postponing a discussion of this topic until now is that I want the reader to see how the essential ideas of Goethean science relate to Steiner's thoughts about modern scientific method.

Johann Wolfgang von Goethe was born in 1749, 22 years after the death of Newton, and died in 1832, a few years after John Dalton calculated the first atomic weights and one year after Charles Darwin boarded the Beagle. The dictionaries tell us that he was a "German poet, dramatist, novelist and scientist whose genius embraced most fields of human endeavor." Goethe considered that his scientific work was no less important than all his other endeavors, so he might not have been pleased to see it left until last on the list. It's generally agreed that he was an outstandingly skillful experimenter and a very acute observer, but most of his contemporaries couldn't stand the way he thought and talked about science, which seemed to them to confuse the artistic and the metaphysical with the scientific. Apart from that pervasive objection, there were two big differences between Goethe's way of doing science and that of his contemporaries. One was that they wanted to do things to nature in order to see what happened, while he wanted to let nature speak for herself; the other is that while physical science was becoming increasingly mathematical, Goethe's work is devoid of mathematics. It may seem odd, therefore, to place a discussion of it immediately after quoting Steiner on the fundamental importance of the mathematical treatment of nature. However…

*

It is generally supposed that Newton's method is analytic and objective, while Goethe's world is poetic and subjective. Rudolf Steiner believed that the meditative approach adopted by Goethe unites the objective and the subjective, enabling different phenomena to be linked together in the human consciousness with the kind of certainty and precision that we usually, although not always correctly, associate with mathematical methods. Goethe's approach is therefore a step towards answering one of the questions asked by Rudolf Steiner in relation to the metamorphosis of scientific method into *spiritual* scientific method.

> ...Could the force which we have to use to attain a mathematical knowledge of nature be used more effectively, with the result not just a mathematical abstraction, but something inwardly, spiritually concrete?

Precision and certainty are not the only desirable qualities; Goethe's attitude is like that of the best mediaeval alchemists – nature will not speak to you unless you approach her with reverence and a pure heart – and we find that Goethe's science is imbued with that kind of moral conscience. It is not only that the scientist must be moral; nature herself has moral qualities. Nature acts and suffers, and is to be loved and cared for, not exploited. Goethe's way of letting nature speak for herself is to study her forms and processes meditatively rather than trying to explain them with conceptual models; he wants to be at one with nature and experience what she does. Now this attitude could easily lead to a lazy, sub-mystical, feel-good wallowing in pleasant sensations; but in spite of his lack of mathematics, Goethe is disciplined, precise and rigorous. His frame of mind is an exact match for the attitude that Steiner describes as necessary for anyone starting on the path of spiritual knowledge.[59]

Goethe did not, however, leave a description of his method – it has to be inferred from his practice. While working as editor of a new edition of Goethe's scientific writings in the 1880's, Steiner realized that something lay behind Goethe's work that was never explicitly stated – not a system of science but a different way of looking at nature – different, that is, from the kind of approach that began before the Renaissance and increasingly became the norm. In ancient times, when people looked at nature they experienced the presence of God, or the gods, and of a varied collection of nature spirits, some mischievous and some benign. Now the divine presence has become much harder to perceive, and the elemental beings have been repelled by concrete and exhaust fumes and by people who think of the land merely as a source of profit, amusement, relaxation, and protection from the less affluent. Even in Newton's time nature still seemed more alive than it does now, and he could refer to mineral processes in organic terms as if a piece of metal had some life in it. We've all heard the story of the apple falling on Newton's head and giving him the idea of gravitation. Chaucer, in the fourteenth century, might have said that the apple fell because the tree had finished with it and because it was time for it to go back to the earth to continue the cycle of regeneration.

> *Every kindly thing that is*
> *Hath a kindly stede ther he*
> *May best in hit conserved be;*
> *Unto which place everything*
> *Through his kindly enclyning*
> *Moveth for to come to.*[60]

Goethe might well have said that the impact damaged Newton's brain and accounted for his subsequent wrong-headedness, although a more ancient commentator might have pointed out that the apple is a symbol of wisdom, and arrived at a different conclusion.

Now, most people believe that the motion of everything from an apple to a planet is determined by the inscrutable action of gravity and inertia. People have given all kinds of answers to the question why there was so little progress in the physical sciences before 1600 and such a tremendous acceleration subsequently. Part of the answer is that instead of seeing natural processes as expressions of something akin to human sympathies and antipathies, the scientists began to treat both mineral and organic matter as if it were dead and subject only to impersonal forces. Gravity makes the apple fall; inertia keeps the planets in their orbits; the sap rises because of surface tension. If nature is inanimate we can do whatever we like to it (no need to say "her" any more) with no twinge of remorse or conscience. Francis Bacon, who was certainly not without a sense of wonder at the marvels of nature, and whose scientific work was done with the ideal of improving the lot of mankind, expressed the modern attitude very aptly, proposing a degree of ruthlessness in dealing with natural objects that a Goethean scientist would find unacceptable.

> *Now as to how my Natural History should be composed, I mean it to be a history not only of Nature free and untrammeled (that is, when she is left to her own course and does her work in her own way)... but much more of nature constrained and vexed; that is to say when by art and the hand of man she is forced out of her natural state and squeezed and moulded.*[61]

It would be hard to think of a plan for the subjugation and exploitation of nature more directly opposed to the kind of contemplative science through which Goethe came to his understanding of plant and animal morphology and the nature of light and color.

*

While working on the first volume of the new Goethe edition, Steiner realized that there was more in it than met the eye, and so he decided that before going on to the second he must explore Goethe's approach to knowledge and try to make explicit what Goethe had left implicit. His study bore fruit in 1886 in the form of *A Theory of Knowledge based on Goethe's World Conception*, a forbidding title for a book that does not make easy reading. You may well feel that Goethe's approach to nature sounds as if it ought to be easier to understand than, say, a regular textbook of physics, but we are talking about two very different ways of thinking. My experience is that Goethean science is easier to do than it is to explain. In a well-written physics text – and I have to interrupt myself here to remark that as a very experienced physics teacher I am aware that that phrase is almost an oxymoron – but, as I was saying, in a well-written physics text, the argument goes from point to point in a logical manner, and anyone of reasonable intelligence can follow it. Now Goethean science is not illogical or a-logical, but it does require a contemplative approach, and you can't get the juice out of it just by reading about it.

The second volume of Goethe's scientific work did not appear until 1887, and another ten years elapsed before the project was completed. One of Steiner's duties as editor was to write an introduction for each section. These were published in English by the Anthroposophic Press in 1950, under the title *Goethe the Scientist*,[62] and the degree of interest which the book stirred up can be gauged by the fact that it was remaindered and in 1964 I was able to buy a copy for $1. *Goethe the Scientist* provides an excellent introduction to Goethean science.

While working on the Goethe edition, Steiner struggled with the nineteenth century version of the problem of knowledge that we have already discussed. In the 1870's, when he began his scientific education, the atom was simply a more sophisticated version of Democritus' indestructible particle. The physicists had had some success with the kinetic theory of matter and the wave theory of light, but there were still only the most rudimentary ideas of chemical combination, and some scientists were still very sceptical about atoms. Steiner, however, could see the way the wind was blowing, and he was particularly distressed by efforts to explain all our perceptions and inner experiences in terms of atomic motion. So, as I mentioned earlier, he found himself faced by the dilemma that had first appeared in the ancient world – if our sense impressions are determined by atomic motions, they must be valueless as a source of truth.

> *It is these reflections that compelled me to reject as impossible every theory of nature which, in principle, extends beyond the domain of the perceived world, and to seek in the sense-world the sole object of consideration for natural science.*[63]

Steiner refers to the battle that he has conducted against the basic conceptions of contemporary natural science, by which he means the view that *"no understanding of the physical world can be gained otherwise than by tracing it back to 'mechanics of the atom,'"*[64] and states the fundamental principle of Goethean science:

> *The theory must be limited to the perceptible and must seek connections within this.*[65]

And as he put it in the *Philosophy of Freedom*, which he wrote while he was working on the Goethe Edition, and which describes his own approach to knowledge:

> *Observation and thinking are the two points of departure for all the spiritual striving of man…*[66]

Twenty-two centuries earlier, to make a true and fruitful reading of nature possible, Aristotle had created a system of logic, and in order to do the same thing at a later stage of human evolution, Goethe, implicitly, and Steiner, explicitly, developed a new form of science. We have the fundamental principle, but what comes next? It's all very well to speak of seeking connections within the perceptible, but how do you actually do this? It is not to be expected that Steiner found, or would have wished to find, any sort of prescribed system of science in Goethe's writings, but he did find two ideas which act as signposts. One is the operation of metamorphosis and the other is the notion of the primal (or archetypal) phenomenon.

*

Steiner points out that Goethe was developing the idea of an entity in which every part animates every other, in which one principle permeates all the details. We read in Faust:

> *How all a single whole doth weave*
> *One in the other works and lives.*

When Goethe refers to a primordial creation and continuing world in which a single being is enduringly manifest through a sea of continuous change, he is wrestling with problems that sound very similar to the ones that had preoccupied the pre-Socratic philosophers 2300 years earlier, but he is doing it as a post-Renaissance man with a highly developed capacity for independent thinking. The same could be said of many nineteenth century scientists who tried to reveal the unity hidden beneath the multiplicity of nature. Goethe, however, came to his scientific work with a different frame of mind, not to be explained, as Steiner says, by simply presenting Goethe as a thinker. In Section VIII of *Goethe the Scientist* Steiner goes into the question of "how this genius came at all to be active in the scientific domain." "Goethe had to suffer much because of the false assumption of his contemporaries that poetic creation and scientific research could not be united in a single mind...." Goethe's artistic nature carried with it a strong inner impulse towards scientific thinking – scientific work was not merely a sideline but a necessary development. "There is an objective transition from art to science, a point where the two come together in such a way that perfection in one field requires perfection in the other."

Most nineteenth century scientists, believing that there was a great gulf fixed between objective science and subjective art, found the conjunction of science and art incomprehensible. This attitude shows up strongly in the course of a talk on Goethe's *Farbenlehre*, given by John Tyndall at the Royal Institution in 1880.[67] Tyndall, a distinguished physicist, a not inconsiderable poet, and a great champion of Goethe the poet and dramatist, said the following:

> *The average reading of the late Mr. Buckle[68] is said to have amounted to three volumes a day. They could not have been volumes like those of the Farbenlehre. For the necessity of halting and pondering over its statements is so frequent, and the difficulty of coming to any undoubted conclusion regarding Goethe's real conceptions is often so great, as to invoke the expenditure of an inordinate amount of time. I cannot even now say with any confidence that I fully realize all the thoughts of Goethe. Many of them are strange to the scientific man. They demand for their interpretation a sympathy beyond that required, or even tolerated, in severe physical research.*

These remarks help us to understand the indifference or hostility with which the *Farbenlehre* (loosely translated as *Color Theory*) was received. Goethe's scientific works demand a meditative reading. One cannot just zip through them as though they were typical science textbooks. Steiner himself found that disentangling the essence of Goethe's message was no easy task. It is likely that I am not alone in finding a similar difficulty with some of Steiner's works. But the struggle is abundantly worth it, for the result is not only the acquisition of knowledge but also the development of capacities. There is nothing like a stiff, self-imposed course of anthroposophy to deepen your perceptions of yourself and the world.

*

For Goethe, "art and science sprang from a single source. Whereas the scientist immerses himself in the depths of reality in order to be able to express its impelling forces in the form of thoughts, the artist seeks by imagination to embody the same forces in his material."[69] Imagination is a power of perception and disciplined artistic creation, not a tendency to arbitrary fantasy.

> '*In the works of man, as in those of nature, what most deserves consideration is the intention,*' *says Goethe. Everywhere he sought, not only what is given to the senses in the external world, but the tendency or intention through which it has come to exist… In nature's own formations she gets 'into specific forms as into a blind alley'; one must go back to what was to have come about if the tendency had been able to unfold without hindrance, just as the mathematician never has in mind this particular triangle, but the nexus of law which is fundamental to every triangle. Not what nature has created, but according to what principle it has created, is the important question. And then this principle is to be worked out as befits its own nature, not as this has occurred in the*

single form subject to a thousand natural contingencies. The artist has to 'evolve the noble out of the common, the beautiful out of the misshapen.' [70]

No crystal, plant, or animal grows freely out of its own inner necessity. Available space, weather, nutrition and the activities of competitors all mask the innate developmental principles. Studies of organic growth are hampered by the great difficulty of distinguishing between the essential and the accidental, but this is exactly what Goethe accomplished in his discovery of the archetypal plant – the plant, that is, that does not merely *represent* all actual plants but has the potential to *become* all plants; the plant which nature, so to speak, had in mind, unblemished by the vicissitudes of its environment.

Goethe believed, and acted on his belief, that artistic vision could reveal those developmental principles. The process is contemplative, not analytical, and requires time and inner quiet. This is in tremendous contrast to both eighteenth century taxonomy and nineteenth century chemistry. The Linnaean system separates nature into compartments whereas Goethe experiences nature as a continuum. Nineteenth century chemistry is the story of people trying one conceptual model after another, in the hope of eventually finding something that will fit an increasingly complex set of experimental observations. In contemplating the forms of plants and animals, Goethe perceived a formative principle of metamorphosis which enabled him to see each organism as a unity of interrelated parts and to perceive principles of growth and being that might apply to the whole process of nature rather than to the restricted fields of the botanist and the zoologist. In the formation of seed, leaves, calyx, corolla, stamens and pistil, fruit, and, again, seed, there is a series of alternating expansions and contractions. Mainstream scientists look for molecular processes to explain the expansions and contractions, but according to Steiner this stands the matter on its head.

> Nothing is to be presupposed which causes the expansion and contraction; on the contrary, everything else is the result of this expansion and contraction. It causes a progressive metamorphosis from stage to stage. People are simply unable to grasp the concept in its very own intuitive form, but demand that it shall be the result of a physical process. **They are able to conceive expansion and contraction only as caused, not as causing.** Goethe does not look upon expansion and contraction as if they were the results of inorganic processes taking place within the plant, but considers them as the manner in which the being of the plant is fully realized.[71]

People who believe that nature consists of nothing but atoms and void, in the highly sophisticated forms which these concepts have acquired in the modern world, feel that we are getting closer and closer to a full explanation of the mechanisms of such processes. I speak with the voice of personal experience when I say that it is very hard, even for those who are intuitively drawn to Goethe's view of nature, to get out of the mechanistic habit. Goethe's way of expressing things has the cognate disadvantages of being easily ridiculed by the scientific *intelligentsia*, and being easily and uncritically accepted by half-baked *dilettanti*, so it is well to remember that the test of a scientific theory is not how good or reasonable it sounds, but how well it fits the facts and, in particular, how fruitful it is in generating further penetration into the mysteries of nature.

Hard as all this may be for our conventional thinking to digest, it actually becomes harder. Steiner continues:

*Nature advances from seed to fruit in a succession of stages, so that the next following appears as a result of the next preceding. This is called by Goethe '**an advancing on a spiritual ladder.**' Speaking about the upwardly growing plant, Goethe says 'that a higher junction, as it arises out of the lower, and receives the sap mediated through this, must receive this still more finely filtered, must also enjoy the influence of the leaves which has been occurring in the interval, must develop itself more finely, and must bring fine sap to its leaves and buds.'*[72]

A nice analogy, one might say, between biological growth and spiritual development. But just as Goethe's view of natural processes was not *merely* scientific, this statement is not *merely* analogical. This is how nature works. Its forms result from organic growth *and* creative intention. This, we might say, is the real meaning of intelligent design: not God creating an elaborate plan and bringing it into full being with a snap – or, perhaps, six snaps – of his fingers, but an ongoing process in which spirit is always working in the physical.

Since the seventeenth century most scientists, if they considered the spiritual at all – and a great many of them did – have been careful to keep it strictly separate from the scientific. For the increasingly atheistic society of the past couple of centuries the problem has largely vanished. But here is a man who speaks of the spiritual and the physical as all part of the same indivisible whole, who links the scientific with the artistic, and blurs the distinction between objective and subjective. He rejects the very bifurcation that made post-Renaissance science possible. No wonder the general opinion was that Goethe ought to have stuck to his poetic last. Yet it is not simply a matter of trying to put the clock back. Metamorphosis is a rule of nature, and if we can calm the inner mechanic long enough we can learn to appreciate its creative operation.

The study of metamorphosis is not going to get the kind of results that we get from mainstream science. We are not dealing with ways of manipulating nature but with ways of seeing, understanding, appreciating and caring for nature, ways, perhaps, of inviting the spirit to return to our woods, fields, shopping malls, and parking lots.

(iii)

The Primal Phenomenon

"Are not the Rays of Light very small bodies emitted from shining substances?" asks Newton, towards the end of his celebrated *Opticks*, published in 1704. "For such bodies will pass through uniform Mediums in right [straight] lines without bending into the shadow, which is the Nature of the Rays of Light." This was Newton's corpuscular theory of light, and it has certain corollaries, one of which concerns the relations of white and colored lights. It was natural for him to suppose that different colors are produced by the action of different particles and that sunlight is a mixture of these particles. He thought that by refracting the particles through different angles the prism separated colors that were already there. A century later, when Goethe was proposing a way of studying nature which rejected mechanistic explanations, Thomas Young and Augustin Fresnel were casting grave doubt on the validity of the corpuscular theory; but several more decades would elapse before the wave theory came into its own, so when Goethe vigorously attacked the view that white light is a mixture of the prismatic colors, it was against Newton that he vented his outrage. Goethe experienced sunlight as a glorious unity, and believed that the prism produced colors by a process of metamorphosis and not by separating the components of a mixture. He therefore found Newton's ideas very offensive.

Goethe's view was that colors are created through the interaction of light and darkness. Darkness seen through light generates cool colors as in the blue of the sky, in which we see the darkness of space through the sunlit atmosphere. Light seen through darkness brings warm colors, as when we see red of the setting sun through the intervening space. Goethe tries to show that these processes operate whether the colors we see are those of the prismatic spectrum, the rainbow or the daylight sky and the sunset, and his work embodies a principle that Steiner was later to make explicit: *the contemplation of such a set of phenomena can lead to the experience of connections that are necessary in just as strong a sense as the connections in a mathematical proof.* Steiner calls such necessary relationships primal phenomena and considers that they are embodied in a new kind of clear and exact thinking that is to Goethean science as mathematical thinking is to orthodox science.

Newton believed that by using a very narrow beam of light he obtained the purest possible colors from his prism; Goethe considered that Newton's experiment, in which the sunlight passed through a narrow slit, was an act of "constraint and vexation," that the resulting spectrum was over-refined, and that Newton's theory of color could only be the product of a sick mind. Using a wide beam of sunlight, Goethe obtained from his prism two regions of color separated by a region of white. The region refracted through a smaller angle consisted of the warm colors, red, orange, and yellow. The region refracted through a greater angle contained blue, indigo, and violet. In thus identifying the colors it is necessary to observe that except at the outer limits of the spectrum their qualities are quite different from those of the colors produced from the narrow beam. Each spectrum has its own kind of beauty. Goethe's colors are full of life, energy, and sparkle, and merge imperceptibly into the white central region. Newton's spectrum is remote and austere. Students in a dark room, seeing the wide Newtonian spectrum produced by a carbon disulphide prism, are apt to gasp with wonder and astonishment. Newton thought that his spectrum was pure, whereas Goethe considered it sterile, and we know that Goethe was not one to tolerate sterility. What you do depends on what you are trying to achieve.

Implicit in Newton's thinking are two assumptions that seem to have become part of the collective post-Renaissance scientific unconscious. One is that nature is a complex of mechanical systems proceeding quite independently of human consciousness but, in principle, perfectly accessible to consciousness. The other is that a system can be analyzed into parts, each of which functions in exactly the same way as part of the whole as it would on its own. Reason and experience gradually convinced the scientific community that these assumptions yield only first approximations to truth. Goethean scientists don't need to be persuaded of this since neither Goethe nor Steiner made these assumptions. Consciousness is the central element in science because it is where the phenomena meet and interact. As the relativists and quantum physicists of the early twentieth century were to realize, the observer is part of the process. This realization produced some pretty sensational new science and some apparently unanswerable questions but left the scientists' orientation towards the natural world very much where it had been before. The *quality* of red as a sensation in consciousness is not something that physicists spend any time on.

*

To a physicist with an orthodox upbringing, the creation of colors by the setting sun has nothing in common with the action of a prism in producing red at the less refracted end of the spectrum. The former is quantitatively understandable in terms of the mathematics of scattering; the latter, with much greater difficulty, in terms of electromagnetic conditions in the glass. The red arc of the rainbow is another kettle of fish with several new ingredients, but it is still susceptible to physical mathematics. To Goethe, Steiner and scientists inspired by their work, red is red whether we see it in a prismatic spectrum, in a rainbow or in a sunset, and the three phenomena are linked by the observation that, in each case, red appears when light is seen through intervening darkness. There is a deep division between the mind-sets of the orthodox and the Goethean scientist. The former wants to explain what happens in terms of molecular, atomic and sub-atomic events; the latter does not wish to *explain* anything, but to understand his perceptions in terms of each other and the circumstances of their production.

It would be hard to think of a better example of the benefits and limitations of the mathematization of nature than the physics of scattering as it was developed in the late nineteenth century. We are familiar with the idea of a beam of light passing in a lawful way through a transparent medium, such as air or glass. If the medium were perfectly transparent, the beam would be invisible from the side. A beam passing through a turbid medium, such as dusty or misty air, is visible from the side because some of the light is redirected by dust particles or droplets of water. *Scattering* is the term used when we have the impression that this redirection takes place more or less at random. Lord Rayleigh, whom we have already met, believed that sunlight, in its passage through the earth's atmosphere, is scattered by minute particles of dust, and he turned this belief into a mathematical formula according to which the degree of scattering varies inversely with the fourth power of the wavelength of the light. To put this in practical terms, we note that the wavelength of light at the red end of the spectrum is about 1.8 times the wavelength at the blue end. According to Rayleigh's mathematics this means that the blue will be scattered more than the red by a factor of 1.8^4, or about 10. There is a strong tendency, therefore, for the red light to pass on through the atmosphere into the beyond, and for the blue to be scattered in all directions, including earthwards, for the delectation of those who are lucky enough to be out and about on fine, sunny days. On their way home in the evening, those fortunate people may be able to see the sunset, at which time they will be looking directly at the sun and will receive light that has lost a great deal of the blue because of scattering, but retains a far greater proportion of the red.[73]

This is all worked out with mathematical precision and fits the observations very well. The big problem about it is that it gives no inkling of the living, inner experience of color. "If one is honest one has to admit that all the knowledge obtained in this way stands as a closed door to the outer world in that it does not allow the essence of this outer world to enter our cognition..." When Goethean science is working properly, its fundamental substance is the quality of the perception – the redness of the red and the blueness of the blue. Such things are absent from modern physics.

Rayleigh's Fourth Power Law eventually found its place in the interlocking system of atomic physics. Einstein showed that dust was not required and that it is the molecules of the air that do the scattering. Rayleigh went on to show that a study of scattering and refraction would yield the information necessary to make an estimate of the famous Avogadro number – the number of molecules in a gram molecule. The number obtained in this way was in good agreement with the results obtained by several other independent methods. Such concordances contributed to a general feeling at the end of the nineteenth century that, in spite of its many problems, the atomic theory was moving in the right direction. Also, by that time, the wave theory of light had completely superseded the corpuscular theory, its tranquil undulations as yet undisturbed by the quantum theory that was just around the corner. Goethe had been outraged by Newton's belief that white light is a mixture of colors, but he might have been even more disturbed by some of the consequences of the wave theory – not only that the difference between the views of sunlight as a mixture of colors and as a glorious unity is merely a matter of mathematical taste and that the two formulations are different ways of saying the same thing, but also that the "glorious unity" is, in a more modern form of the theory, the manifestation of a stream of separate, randomly emitted pulses.

*

The point that needs continually to be brought to mind is that to Goethe and Rudolf Steiner red is red whether we see it in a prismatic spectrum, in a rainbow or in a sunset. What Goethe saw as the basic principle is that colors appear though the interaction of light and dark. The contemplation of such phenomena leads to the experience of connections between events in the natural world that seemed quite disparate to the orthodox mind. Like metamorphosis in the vegetable world, the interaction of light and dark needs no explanation. This being the case, Goethe regarded it as a primal phenomenon and valued the observation because it enabled him to gather together phenomena which could be understood as explaining each other.

Goethe's work in optics has its problematical aspects even among anthroposophists, and Steiner stated explicitly that he had no desire to defend every detail of the *Farbenlehre*. It is therefore important to realize that Goethean science is not a collection of observations and theories but a disciplined way of seeing and understanding, developed from the perception of Goethe's attitude to nature and his effort to raise the science of qualitative experience to the level of precision and clarity found in mathematical science. This approach can bring us closer to the natural world and increase the intensity with which we experience its tastes, smells, colors, sounds and textures. In the process it may also help us to become better farmers, gardeners, physicians, artists and teachers.

*

If we are wondering whether a commitment to Goethean science is compatible with the acceptance of any form of atomic theory, we are not likely to get much comfort from Rudolf Steiner. His last great public scientific course ended on January 6, 1923, but he continued to share his thinking about the problems of the atom with members of the Anthroposophical Society and gave one of his most trenchant, radical and disturbing lectures three weeks later. During the thirty years between Steiner's early refutations of atomic theory and this late challenge to conventional thinking, the atom had been transformed into something that none of the nineteenth century theorists would have recognized, and everyday life had become permeated with electrical technology. In the next section I give some more of the exoteric background for Steiner's extraordinary perceptions of what happens when our thinking is increasingly permeated by the electrical and mechanical concepts that Niels Bohr used in creating his model of the atom.

The evolution of the Bohr atom is historically just as complex and easily misrepresented as the emergence of Planck's quantum. The logic behind it can be expressed in a straightforward way but, as usual, history does not conform to logic. I tell as much of the tale as I think will be helpful to the serious reader, starting with that great monument to the industry and perspicacity of the nineteenth century, the Periodic Table. In the enormous amount of labor involved in creating it and its inscrutability with regard to meaning and purpose, the Table somewhat resembles Stonehenge. One of the great achievements of the Bohr atom was that it gave an approach to unraveling its secrets.

My purpose in the chapter that follows is to show that, like Planck's quantum, Bohr's atom was the result, not of mental inertia, but of an intense striving to find a unifying explanation for a wealth of apparently unrelated observations.

Chapter IV
Bohr's Atom – Antecedents

(i)

Periodic Tables

In Chapter I, Section (vi), I spoke of John Dalton's attempt to find the atomic weights of the elements. The enormous difficulty of this project can be illustrated by the apparently simple example of oxygen. The question was, "How many times heavier is one atom of oxygen than one atom of hydrogen?"[74] Now we all know (or think we know) that the formula for water is H_2O – two atoms of hydrogen to one of oxygen – but Dalton didn't know this. We also know that in water, 8 grams of oxygen are combined with 1 gram of hydrogen. Dalton's analyses were not very accurate, but we'll take the accepted value here. In the absence of any further knowledge, Dalton simply assumed that the formula for water is HO – one atom of hydrogen per atom of oxygen – from which it followed immediately that the atomic weight of oxygen must be 8. It took a great deal of work on combining volumes of gases by the great French scientist Gay-Lussac and the Italian Avogadro to make it very likely that the correct formula is H_2O, so that the atomic weight of oxygen must be 16 instead of 8, and I'm not sure that Dalton was ever quite convinced. Be that as it may, my point is that this is only a very simple example of a problem that troubled chemists throughout the first two thirds of the nineteenth century, namely that if you wanted to be sure of a chemical formula, you had to know the atomic weights of the elements involved, and that if you wanted to know the atomic weights you needed to be quite sure of the formula of at least one compound containing these elements. Meanwhile, the number of known elements was increasing rapidly from Lavoisier's thirty of 1789 to over fifty by 1860, and their physical and chemical properties were enthusiastically investigated.

Most of the elements known to modern chemists are never found alone in the wild. Many are extremely active and very dangerous. The alkali metals, such as sodium and potassium, react with air and water spontaneously and at times explosively, producing very caustic compounds. Of the halogens, fluorine is the deadliest gas known, destroying organic tissue on contact; chlorine is famous for its use as a poison gas in the Great War; bromine is a noxious, oily brown liquid; and iodine, a purple crystalline substance, is socially acceptable because, in small quantities, it kills bacteria instead of people. The alkali metals take the chemical properties of traditional metals, such as copper or iron, to an extreme; the halogens take the properties of nonmetals to the opposite extreme. As often happens in political life, the two extremes have something in common – violent and dangerous activity. Between the extremes there are well-behaved elements, unexpected elements and weird elements. Zinc, for instance, is a well-behaved metal; it is shiny, dense, malleable and ductile. It is quite a good conductor of heat and of electricity. It liberates hydrogen from dilute acids, forming salts such as zinc sulphate. It is a perfect, textbook example. It also doesn't exist free in nature. It may seem odd to say that iron is an unexpected element; everyone is familiar with it and its chemical behavior is reasonably satisfactory, although much less straightforward than that of zinc. A visitor from another world, making a survey of the metals, would find, however, that the magnetic properties of iron came as a complete and unique surprise. Its behavior around its melting point is almost as surprising. Like water and quartz it contracts when you would expect it to expand. The different crystalline structures produced by varying forms of heat treatment give us the possibility not only of swords, ploughshares and horseshoes, but also of cast-iron forms and wrought-iron gates. This display of versatility depends also on the ability of iron to dissolve small quantities of carbon, creating steel. Carbon is another story; it is possibly the weirdest of the elements. It may appear as diamond, the hardest known natural substance, or as graphite, one of the best lubricants. It may appear to be quite amorphous, as charcoal. It burns, forming a weakly acidic oxide and making a somewhat tepid pretense of being a normal nonmetal. It is light and neither malleable nor ductile, but it resembles a metal insofar as it is a good conductor of electricity. Unlike well-behaved metals, it has the strange property that its ability to conduct electricity increases as its temperature rises. Unlike any other substance, it forms thousands of different compounds with hydrogen alone and many more thousands with various combinations of hydrogen, oxygen and nitrogen. The elements, whether old, like copper, silver and gold, or new, like sodium, calcium and fluorine, to name just a few, have a tendency to settle in the mind like characters in a story, with personalities, likes and dislikes. I am not sure whether or not this is a good thing, but it is what made chemistry fascinating to me when

I was quite a small boy, and made learning an enthralling experience.

The nineteenth-century chemists observed that these characters tend to live together in families. In addition to the alkali metals and the halogens there is a well-defined family of alkaline earth metals, of which calcium and magnesium are the best-known examples, and an equally clear family of sulphur-like nonmetals. Copper, silver and gold, and zinc, cadmium and mercury form menages a trios, and there is an extended family of useful metals, including iron, nickel, cobalt and platinum. It is to be emphasized that these groupings are based on similarities in chemical and physical properties and are quite independent of any knowledge of atomic weights, atomic numbers and electrons. Some of these properties are obvious ones, like melting and boiling points, density, electrical conductivity and the alkalinity or acidity of the oxides. Others are more subtle, such as similarities in the crystalline forms of compounds such as the chlorides of sodium and potassium.

Chemical behavior was, in fact, even more important than physical properties in establishing groups. All the alkali metals react spontaneously with water at room temperature, liberating hydrogen and forming hydroxides which dissolve and form strongly alkaline solutions. With the exception of lithium, the reaction is so vigorous that the metal melts and fizzes around on the surface, sometimes igniting the hydrogen in the process. A little red litmus solution, added to the water, rapidly turns blue. The alkaline earth metals are much more subdued in their reactions with water, forming hydroxides which are sparingly soluble and only weakly alkaline. No other metals known in the first half of the nineteenth century react in this way.[75] Nearly all the salts of the alkali metals – carbonates, chlorides, sulphates, nitrates – dissolve readily in water. The carbonates of all other metals do not. Apart from all these physical and chemical properties, each of the families we are discussing is united by the possession of a common characteristic that chemists came to regard as of fundamental importance, namely *valence*, and this is where the knowledge of formulae and atomic weights became important.

To the nineteenth century chemists, who had gone through enormous difficulties in finding the chemical formulae of even the simplest compounds, *valence* meant the number of atoms of hydrogen that one atom of an element would combine with or displace. All the halogens react with hydrogen to form acids in which one atom of the halogen combines with one atom of hydrogen, so the halogens are all univalent. All the alkali metals react with the halogens to form salts in which there is one atom of the halogen per atom of the metal – as in common salt, sodium chloride – so, by transference, so to speak, the alkali metals must all be univalent. The formula for water being H_2O, oxygen must be bivalent and the same applies to sulphur. All the alkaline earth metals (the calcium family) are bivalent, and there is also a family of trivalent elements, including boron and aluminum. By the mid-1860's many of these relationships had been established, but no one had any clue about the mechanism of valence. Finding the correct valences depended on knowing the correct chemical formulae, which, in turn, depended on knowing the correct atomic weights of the elements. This sounds logical but, as we have seen, the logic works both ways; finding the correct atomic weights depended on knowing the correct formulae. This, at any rate, was one point at which these family groupings made serious contact with the notion of atomic weight, for by this time the atomic weights of nearly all the elements had been agreed upon and were known with fair accuracy, by which I mean within about 0.1% of the values accepted in the early twentieth century.

While all this quantitative information was satisfactory and useful, it was frustrating because it shed no light on the nature of the atoms, their physical properties or the mechanisms by which they combine to form molecules. A suspicion had already arisen, however, that a careful examination of the atomic weights of elements within a given family might reveal some form of arithmetical regularity. As early as 1817 and 1829, Johann Wolfgang Dobereiner (1780-1849) had observed several groups of three chemically similar elements in which the atomic weight of the middle one was roughly the mean of the other two. One such group was chlorine (35.5), bromine (80) and iodine (127). Dobereiner not only shared Goethe's Christian names, he was also Goethe's chemistry teacher for a while, and it is tempting to surmise that the idea of metamorphosis inherent in the grouping might owe something to his student. Looking for such relationships became quite a common pastime among chemists. In 1843 Leopold Gmelin reported several more such triads and stated his belief that such things were manifestations of some kind of internal atomic structure. William Odling, who had already made valuable contributions to the evolution of organic chemistry, noted a rather fanciful trend in the numerology of atomic weights and pointed out that such relationships were of no value *unless the elements concerned were phenomenologically linked.* In 1857 he produced a set of thirteen groups based on physical and chemical properties – a kind of table in which atomic weights appeared only indirectly, but which anticipated many of the groupings in later tables.

In the 1860's the search produced some even more striking results. Between 1863 and 1866, J. A. R. Newlands, a London industrial chemist, published a series of papers in which he arranged the elements in ascending order of atomic weight, observing that every succeeding eighth element was "a kind of repetition of the first." This worked extremely well for the first seventeen elements. The lithium (Li), sodium (Na), potassium (K) family and the beryllium (Be), magnesium (Mg), calcium (Ca) family come together very nicely in the second and third columns, and the less obvious grouping of the tetravalent elements carbon (C), silicon (Si) and titanium (Ti) in column five, is satisfactory. It also brought two of the halogens, fluorine (F) and chlorine (Cl) together, although it put them in the same column as hydrogen. The equally appropriate vertical pairings of boron (B) and aluminum (Al), nitrogen (N) and phosphorus (P), and oxygen (O) and sulphur (S) are compromised by the seemingly unsuitable company of chromium (Cr), manganese (Mn) and iron (Fe).

H	Li	Be	B	C	N	O
1	2	3	4	5	6	7
F	Na	Mg	Al	Si	P	S
8	9	10	11	12	13	14
Cl	K	Ca	Cr	Ti	Mn	Fe
15	16	17	18	19	20	21

Newlands was the first to use the term atomic number for the ordinal of each element in the sequence. In his table the first three alkali metals, lithium, sodium and potassium, are numbers 2, 9, and 16; the first three alkaline earth metals, beryllium, magnesium, and calcium are numbers 3, 10, and 17. "Members of the same group of elements," he said, "stand to each other in the same relation as the extremities of one or more octaves in music. This peculiar relationship I propose provisionally to term the Law of Octaves." Sceptics made the musical reference an object of ridicule, while the combined effects of missing elements, a few inaccurate atomic weights and the complex chemistry of the heavier elements produced enough "noise" in his system to provide them with reasons for rejecting the idea. One of them derisively asked whether Newlands had tried arranging the elements in alphabetical order. It was not until the 1880's, by which time Mendeleev had been awarded the Royal Society's Davy medal for his periodic tables, that the work of Newlands gained any official recognition. Twenty-one years after reading his major paper to the Royal Society, he too was awarded the Davy medal for his discovery, but only after jogging the Society's collective memory by republishing his papers in book form. Meanwhile, Odling, who in 1865 had published a table very similar to Mendeleev's periodic table of 1869, seems to have been completely forgotten.

(ii)
From Siberia with Love

Dmitri Ivanovich Mendeleev, the fourteenth[76] and youngest child of a teacher, was born in Tobolsk, Siberia, in 1834.

That the name of Mendeleev is engraved on the hearts of all chemists is largely due to the devotion of his mother, who took full responsibility for the family after his father became blind. When Mendeleev was fourteen his father died and the glass factory which had provided the family's livelihood was destroyed by fire. His mother took him two thousand miles to Moscow, only to find that Siberians were not admitted to the university. A further trip to St. Petersburg used up the last of the family's resources and Mendeleev entered a teacher training college there in 1850. His mother died a little later in the year, and we are not told what happened to the other thirteen children. Dmitri Ivanovich, at least, was grateful. "She instructed by example, corrected with love, and in order to devote her son to science left Siberia with him, spending her last resources and strength."

Mendeleev arrived independently at his version of the periodic table in Russia in 1869, while holding a professorship at the University of St. Petersburg. Later the same year, a table embodying exactly the same principle was created by Lothar Meyer in Germany. It was Meyer's publication of an early form of his table, complete with an acknowledgement of Mendeleev's precedence, that helped to put the Siberian's name prominently before the scientific community. The following chart shows the first seventeen elements as given by both Newlands and Mendeleev in ascending order of atomic weight, but with the more modern terminology of groups and periods as given by Mendeleev and Meyer in their later tables. The position of hydrogen remained obscure for a while. Atomic numbers are printed in italics above the names of the elements, and atomic weights below.

Bohr's Atom-Antecedents — 115

Group	I	II	III	IV	V	VI	VII
Period 1	*1* Hydrogen 1						
Period 2	*2* Lithium 7	*3* Beryllium 9.4	*4* Boron 11	*5* Carbon 12	*6* Nitrogen 14	*7* Oxygen 16	*8* Fluorine 19
Period 3	*9* Sodium 23	*10* Magnesium 24.3	*11* Aluminum 27	*12* Silicon 28	*13* Phosphorus 31	*14* Sulphur 32	*15* Chlorine 35.5
Period 4	*16* Potassium 39	*17* Calcium 40					

Starting with hydrogen, the element with the smallest atomic weight, we go across the page, listing the elements in ascending order of atomic weight. As soon as we arrive at an element resembling one already listed, we go back to the left and start a new line. The horizontal lines are called *periods* and the vertical columns *groups*. Except for the placing of hydrogen, which resembles the alkali metals in being univalent, Mendeleev's first seventeen elements form exactly the same groups as those of Newlands. No wonder Newlands was upset! The second period requires seven steps to get from lithium to a metal with very similar properties, sodium. Another seven steps bring us to potassium, which clearly belongs with lithium and sodium, and one more step puts calcium where it belongs, under magnesium. What happens next is much less clear. In Newlands's table the next element was chromium, but the atomic weight of chromium had been corrected by the time Mendeleev put his table together. The next available element in Mendeleev's atomic weight sequence was titanium. Titanium is not a particularly well-behaved metal. Unlike the very proper zinc, for example, it has a hard time producing recognizable salts and seems more comfortable mimicking nonmetals by producing titanates, salts of the very dimly existent titanic acid, which, like the ill-fated liner of the same name, has a difficult relationship with water. It is in the form of titanates that titanium is frequently found in nature, often associated with silicate minerals. It resembles silicon so much, in fact, that it clearly belongs in Group IV, which is where Newlands had placed it, under silicon, rather than in Group III, under aluminum, which is unambiguously metallic. Mendeleev therefore decided that titanium must be given the number 19, and filled the space under aluminum with a question mark, concluding that the real number 18 was out there somewhere and had so far evaded captivity. This was only the start of his problems. After titanium he found himself faced with a row of eight metals – vanadium, chromium, manganese, iron, cobalt, nickel, copper and zinc – none of which seemed to bear a very striking resemblance to anything that had gone before. After zinc, the next available element was arsenic, an element closely related to phosphorus in its chemical properties and clearly belonging in Group V. This meant that Mendeleev had to leave two more empty spaces and accommodate the intervening elements by dividing his vertical groups into subgroups. This is clearly shown in the Periodic Table shown on page 123. As we shall see in the next chapter, one of Bohr's greatest successes was that his atomic model made some sense of this profusion of metals.

(iii)
Predictions, Confusions, Rare Earths, and Hafnium

The periodic tables of Mendeleev and Meyer were received with much greater interest and courtesy than that of Newlands, but what really fired the imagination of the chemists was the outcome of certain predictions that Mendeleev made. Leaving spaces for elements in Groups III and IV was a tacit suggestion that one day elements for these slots would be discovered. Mendeleev went further, however. He predicted not only the discovery of the missing elements but also their atomic weights, their specific gravities and the formulae of some of their compounds. In 1875, Emil Lecoq de Boisbaudran, who was at the time in a state of blissful ignorance about the periodic table, isolated Mendeleev's missing element number 28, but it was a long time before he realized what he had done. By the mid-1860's, thanks to Bunsen and Kirchhoff, spectroscopy had become a valuable tool in chemical analysis, and that is how Lecoq de Boisbaudran got in on the act. Characteristic lines in the spectra of light from specimens of minerals held in the non-luminous flame of a Bunsen burner make it possible to identify some of the elements without actually isolating them. Other elements require the use of an electric arc or spark. In 1875 Lecoq, working with zinc blende from the Pyrenees, discovered the spectrum of an unknown element which, as a patriotic Frenchman, he named Gallium. He then obtained enough of the metal to measure its atomic weight as 69.9 and its specific gravity as 4.7. Now Mendeleev had predicted that his element number 28 would have an atomic weight of 68 and a specific gravity of 6.0. Hearing about Lecoq's discovery, he concluded that Gallium was indeed number 28 and that Lecoq was mistaken about the density. Lecoq took Mendeleev's conclusion as a claim of priority and an insult to his (Lecoq's) native country. When the specific gravity of gallium was measured again, however, it was found to be 5.9, very close indeed to Mendeleev's prediction. Chemists were amazed to find that Mendeleev seemed to know more about the element than the man who discovered it. The apparently omniscient Russian had further successes with the discoveries of elements scandium and germanium, which fitted his specifications with impressive accuracy. There were still some missing elements, however, and eventually the search for one of them would result in a very important event in the life of Niels Bohr.

*

Having for a long time refused to admit the existence of the element needed for Group III in the fourth period of the table, nature seems to have made up for this obstinacy by providing no fewer than fifteen elements for the Group III space next to barium in the sixth period. Metallic oxide ores were known as earths, and in 1794 John Gadolin discovered a new earth, although unfortunately not a new heaven, in the mineral gadolinite which now bears his name. A full account of the unearthing, so to speak, of the fifteen elements may tend to induce a feeling of repletion, so I have made it as brief as possible. It has its point, as we shall see later.

Gadolinite is a silicate mineral found near the town of Ytterby in Sweden, so its newly discovered component was called *yttria*. It was not until 1843 that C. G. Mosander found that yttria is a mixture of different metallic oxides, three of which he identified as the oxides of unknown metals which he named *yttrium, erbium,* and *terbium.* Ytterby may have been an obscure little township, but it now had the distinction of having three elements named after it, and that wasn't the end. Yttria continued to yield subtly differing components.

In 1878, Marignac found the oxide of a fourth metal, which he called *Ytterbium*. A couple of years later P. T. Cleve, having discovered two new metals and evidently thinking that Ytterby had been exhausted as far as the provision of names for new elements was concerned, christened one of them *Thulium*, Thule being the ancient name for the most northerly habitable region of the earth, and the other one *holmium*, after Stockholm. A few years later Marignac found *gadolinium*, closely associated with terbium.[77] That makes seven different metallic oxides all found in one mineral, and it still wasn't the end. The oxides were so similar that it took chemists over a century to disentangle them all, the ninth and last one, *lutecium*,[78] being discovered by G. Urbain in 1905. Lecoq continued with his spectroscopic and chemical work, sharing with Marignac the credit for the discovery of gadolinium, and identifying *samarium and dysprosium*. Dysprosium, the eighth of the elements from yttria, had been cunningly concealed in the earth in which holmium had been found, and it got its name from the Greek word for "difficult." Samarium came from a different source.

While all this was going on, another sequence of discoveries resulted from the identification in 1803 of an earth called *ceria*, by Martin Klaproth, the first professor of chemistry in the University of Berlin. The great Swedish chemist Berzelius, who had discovered ceria independently, was convinced that both it and yttria were mixtures of several different oxides, and he persuaded Mosander to try to analyze them. Samarium came from a mineral called samarskite, after the Russian M. Samarsky, in which Mosander had found a mineral very much resembling ceria, from which the metal cerium had been isolated. From the samarskite ceria he had isolated the oxide of a previously unknown metal which he called *lanthanum*, from the Greek verb meaning *to hide*. Coupled with lanthanum he found *didymium*, named from the Greek word meaning a twin, and it was from didymium that Lecoq separated samarium. As with the yttria sequence, new elements continued to pop up. *Praseodymium* (leek green, from the color of its oxide) and *neodymium* (new) were separated from didymium by Auer von Welsbach, the inventor of the incandescent gas mantel. Europium was separated from samarium by Demarçay in 1901, and, finally, *illinium* was detected among the lanthanum elements in 1926 by A. J. Hopkins at the University of Illinois.

Of the metals mentioned in the preceding paragraphs, yttrium already had a place in the periodic table in Group III next to strontium. This leaves fifteen elements, known collectively as the rare earth elements, all metals and all trivalent,[79] with atomic weights ranging from 138.9 (lanthanum) to 175 (lutecium). Lanthanum, having been discovered in 1839, seemed to be a perfectly respectable element and was well enough known to be placed quite confidently below yttrium in Group IIIa, next to barium. The element after lutecium in the atomic weight series is hafnium, an element that managed to conceal its presence so well that its existence wasn't suspected until 1911. It was finally tracked down in 1923 hiding in the minerals of zirconium, a metal which it so greatly resembles that the two could be separated only by a tedious process of fractional crystallization of obscure complex ammonium salts. With its valence of 4 and its similarity to zirconium it fits very well in Group IV, providing a kind of bookend for the rare earth elements, but in the 1870's Mendeleev and other chemists who were trying to figure out how to fit this proliferation of almost identical metals[80] into the periodic scheme, didn't know that. The next available element at that time was tantalum, a pentavalent metal very much like vanadium and niobium, and quite at home in Group V. Mendeleev believed that more detailed research would show that the rare earth elements constituted a whole period of the table that would fit between lanthanum and tantalum, but the situation remained in flux until after Mendeleev's death in 1907. Increased familiarity with the properties of the rare earth metals, coupled with the discovery of hafnium and the new theories of atomic structure, eventually made it clear that all the elements from lanthanum to lutecium really belong together in Group III(a). In the year 2012, when we think we understand exactly how the system works, it is easy to say that in the context of the periodic table *as a whole*, the intrusion of fourteen superfluous, almost identical elements was inexplicable. There is some danger of forgetting that for fifty years after Newlands published the earliest version, the periodic table as a whole was inexplicable – a phenomenological marvel that kept its inner secrets well hidden.

In 1879 Lars Nilson discovered a new metal while working on – of all things – yttria. At first it seemed to be just another of the pack of rare earth elements, but it turned out that its atomic weight was 43.8, its specific gravity was 3.86, and it formed an oxide X_2O_3. It didn't take long for people to remember that Mendeleev needed an element to go between calcium and titanium in the third period of his table, and that he had predicted that the element would have an atomic weight of 44, a specific gravity of 3.5 and would form an oxide X_2O_3. Nilson patriotically called the new element scandium, and it took its rightful place in the table with little opposition. Seven years later Clemens Winkler, a German chemist, discovered the element that Mendeleev had predicted for Group IV. Naturally he called it germanium. Mendeleev had its atomic weight and specific gravity right within ½ of 1% and had correctly predicted the formulae of its oxide and chloride.

No matter how confused the situation became with regard to the rare earth elements, the periodic table was now installed as an accepted, even dominating, feature of the chemical landscape and a version of it as it might have appeared in the 1920's is printed on the following page (*Table 1*). By that time the inert gases had been discovered, constituting a whole new group, usually designated as Group 0, but leaving the essential nature of the table unchanged. It wasn't until 1913, that any plausible explanation of its structure began to appear, and part of the input for this new beginning came from an obscure Swiss schoolmaster called Balmer.

(iv)

The Hydrogen Spectrum

Johann Jacob Balmer (1825-1898) had a Ph. D. in mathematics, taught math and Latin at a girls' school in Basle, and made a serious hobby of spectroscopy. Like many other people at the time, he was fascinated by the light emitted when electricity was passed through rarefied gases in discharge tubes. When light from such a source is passed through a prism or a diffraction grating, the result is not a continuous spectrum like that of sunlight but a set of separate lines. In the case of hydrogen there is an intriguing set of five lines that appear to have some pattern, and it was Balmer who, in 1884, gave the pattern mathematical expression.

Periodic Table – Elements Known in the 1920's

Period	Group I a	Group I b	Group II a	Group II b	Group III a	Group III b	Group IV a	Group IV b	Group V A	Group V b	Group VI a	Group VI b	Group VII a	Group VII b	Group VIII	Group 0
1	*1* Hydrogen 1.008															*2* Helium 4
2	*3* Lithium 6.94		*4* Beryllium 9.02		*5* Boron 10.8		*6* Carbon 12		*7* Nitrogen 14		*8* Oxygen 16		*9* Fluorine 19			*10* Neon 20.2
3	*11* Sodium 23		*12* Magnesium 24.3		*13* Aluminum 27		*14* Silicon 28		*15* Phosphorus 31		*16* Sulphur 32		*17* Chlorine 35.5			*18* Argon 39.9
4	*19* Potassium 39.1	*29* Copper 63.6	*20* Calcium 40	*30* Zinc 65.4	*21* Scandium 45.1	*31* Gallium 69.7	*22* Titanium 48	*32* Germanium 72.6	*23* Vanadium 51	*33* Arsenic 75	*24* Chromium 52	*34* Selenium 79	*25* Manganese 55	*35* Bromine 80	26 Iron 55.8 27 Cobalt 59 28 Nickel 58.7	*36* Krypton 83.7
5	*37* Rubidium 85.4	*47* Silver 107.9	*38* Strontium 87.6	*48* Cadmium 112.4	*39* Yttrium 88.9	*49* Indium 114.8	*40* Zirconium 91.2	*50* Tin 116.7	*41* Niobium 92.9	*51* Antimony 121.8	*42* Molybdenum 96	*52* Tellurium 127.6	*43* – (99)	*53* Iodine 126.9	44 Ruthenium 102 45 Rhodium 104 46 Palladium 107	*54* Xenon 131.3
6	*55* Caesium 132.9	*79* Gold 197.2	*56* Barium 137.4	*80* Mercury 200.6	*57-71* Lanthanum - Lutecium 139-175	*81* Thallium 204.4	*72* Hafnium 178.6	*82* Lead 207.2	*73* Tantalum 180.9	*83* Bismuth 209	*74* Tungsten 183.9	*84* Polonium 210	*75* Rhenium 186.9	*85* – –	76 Osmium 190.2 77 Iridium 193.1 78 Platinum 195.2	*86* Radon 222
7	*87* – –		*88* Radium 226		*89* Actinium 227		*90* Thorium 232.1		*91* Protoactinium 231		*92* Uranium 238.1					

Table 1

At this time, wave theory of light was in the ascendant, and there was no difficulty in calculating the wavelengths corresponding to the spectral lines. Looking at the spacing of the lines, Balmer had the feeling that they must be governed by some kind of mathematical law, and eventually he showed that the wavelengths fit the following formula, which I quote for people who appreciate such things:

$$1/\lambda = R(1/n^2 - 1/m^2)$$

Lambda (λ) is the Greek letter traditionally used by physicists to represent wavelength. In the Balmer Series, $n = 2$ and m is any whole number from 3 to 7. R is a mysterious number known as the Rydberg constant, after the Swedish physicist who labored over vast quantities of spectroscopic data and put Balmer's formula into a more general context. *The important point to note is that the wavelengths of the light emitted by agitated hydrogen depend on the values of two small whole numbers.*

Variables that can take only whole-number values were almost unheard of in physical science[81] and Balmer must have been intrigued by their presence. He wondered why n had to be 2 and conjectured that there would be other sets of lines in the hydrogen spectrum corresponding to $n = 1; 3$ and perhaps higher integers. He turned out to be right, but this wasn't confirmed experimentally until sixteen years after his death. Rydberg believed that the spectrographic evidence might contain clues to the nature of the Periodic Table, but no one came to any useful insight into the meaning of it all until Bohr came along.

(v)
Cathode Rays

In Chapter I, Section (x), I mentioned the research on the conduction of electricity through rarefied gases that led to J. J. Thomson's discovery of the electron. Now, in preparation for the story of Bohr's atomic model, we must look at this work and the contemporary discovery of radioactivity in more detail.

Once systematic investigation began, it became clear that the pressure of the air was a key variable in deciding what happened inside the discharge tube. A new generation of vacuum pumps, invented in the 1850's, enabled workers to reach much lower pressures. Extraordinary displays of color were seen in tubes much smaller than the one used by Watson, and it was shown that at very low pressure air becomes a little more like a traditional conductor of electricity, although there are still great differences. Ohm's Law, for instance, is not obeyed; doubling the electromotive force does not double the current as it does with metallic conductors.

Even at low pressures air seems not to become a *good* conductor, since an electromotive force of a thousand volts or so is needed to produce a convincing display in a tube 25 centimetres long, but at least the current flows in an orderly, continuous way and can be measured with a galvanometer. The numbers – the electromotive force (volts), current (ampères) and the air pressure – were of the greatest interest to the physicists, and it is amazing to see how much about the internal organization of a gaseous discharge was deduced with the aid of accurate pressure gauges, galvanometers and adjustable electric and magnetic fields; but it was the displays of color in the tube that caught people's imagination most strongly.

If we start with air at normal atmospheric pressure and gradually evacuate the tube, a long, broad and somewhat stabilized spark, mostly purple but flecked with yellow, appears. This is sometimes referred to as the *broad spark*, and it is what Watson had observed in 1752. As the pressure decreases below one five hundredth of an atmosphere, the broad spark is transformed into a stable pink column occupying most of the space between the anode (positive electrode) and the cathode (negative electrode) but leaving a small gap next to the cathode. This is known as the *positive column*. The gap is called *Faraday's dark space*, and a purple glow which appears on the cathode at the same time is known as the *negative glow*. As the pressure continues to diminish, the positive column shortens and breaks up into striations. Meanwhile, the negative glow moves away from the cathode, leaving another gap (*Crookes dark space*) and another glow appears on the cathode (the *cathode glow*).

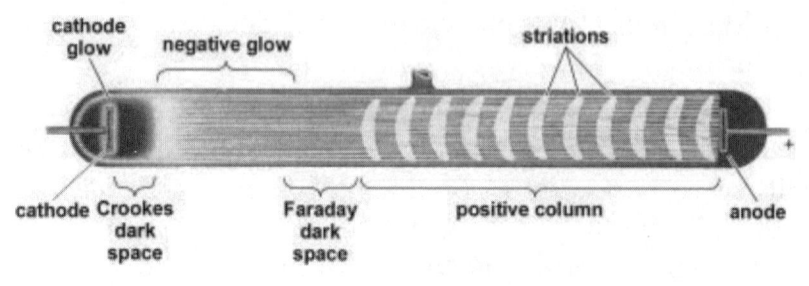

Figure 3

Further diminution of the pressure banishes almost all color from the inside of the tube, leaving only the cathode glow and a faint violet glow throughout the tube. One gets the impression that the Crookes dark space has filled the whole tube. At this point the glass starts to fluoresce with a ghostly green light.

When Julius Plücker (1801-1868), a great mathematician who became professor of physics at the University of Bonn, reached this stage in about 1858 it was not only because of the improvements in vacuum pumps and the invention of the induction coil, but also through the cooperation of a highly skilled technician, Johann Geissler. Geissler had set up shop in Bonn in 1852 and was ready to provide Plücker and his colleagues with all the instruments that a nineteenth century science faculty could desire, and perhaps some that they wouldn't have thought of without him. He was also an excellent glassblower and was able to provide vacuum tubes with the electrodes sealed directly into the glass. This technique was a great improvement on the use of leaky rubber plugs to hold the electrodes in place. Using these smaller, manageable tubes and his convenient, reliable induction coil and vacuum pump, Plücker was able to describe in detail all the stages of electrical discharge as the pressure decreased to about one thousandth of an atmosphere. Having reached the lowest pressure possible with his apparatus, he placed the tube between the poles of a powerful horseshoe electromagnet and found that the violet glow moved in an arc towards the wall of the tube, where it appeared to produce a concentrated area of green fluorescence.

It is perhaps unnecessary to say that these phenomena were quite astonishing and at first perfectly unintelligible to those who discovered and studied them. Clearly some form of energy was passing through the long space between the electrodes, but the question of how or in what form was not to be answered without a great deal of further research. Plücker observed that the arc of violet light behaved as if it were constructed of tiny magnets. Investigating these phenomena in the 1870's and '80's, Sir William Crookes (1832-1919), an accomplished experimental physicist, became convinced that he was witnessing the manifestations of a fourth state of matter, and spoke eloquently about having reached the borderland between the known and the unknown, where he expected that the greatest scientific problems of the future would find their solution.[82] Some physicists believed that the electrical forces were breaking up the molecules of the air and producing charged particles resembling the ions credited by the great Swedish chemist Svante Arrhenius (1859-1927) with carrying electricity across aqueous solutions in electrolysis.

There was great interest in the final stage of evacuation, when all the more striking color displays had vanished and the green fluorescence was most prominent. Tubes of all shapes and sizes were constructed. Some were shaped like an early television tube so that anything emitted from the electrode at the narrow end would strike the glass at the wide end directly, causing the usual green fluorescence. In 1869 Johann Wilhelm Hittorf (1824-1914), a student and research associate of Plücker, discovered that a metal object placed in such a tube cast a geometrical shadow in the fluorescence, provided that the electrode at the narrow end was the negative. Seven years later, Eugen Goldstein (1850-1930) went on to show that this was still the case even if the cathode was a plane sheet rather than something approximating a point.

It seemed therefore that electrical rays must be emitted from the cathode at right angles, and travel in straight lines along the tube. The direction in which the rays were deflected by a magnet indicated that if they consisted of particles, the particles must be negatively charged. This made sense since the particles, if such they were, seemed to be emitted at great speed from the negative electrode. Goldstein, however, believing that he was dealing with some sort of wave motion, called the emissions *cathode rays* (*Kathodenstrahlen*). He was in the minority in this belief, but the name stuck. Crookes found that a delicately balanced paddle wheel placed in the tube rotated in the sense that would be expected if the rays traveled from negative to positive. This seemed to show that the rays carry some momentum – that is to say, they would exert a force on anything that got in the way. Some even thought that the motion of the paddle wheel proved that the rays must consist of material particles, forgetting, apparently, that waves carry momentum and exert pressure. J. J. Thomson, however, showed that the momentum of the rays would not suffice to turn the wheel and that the motion was caused by a rise in temperature on the side of the vane on which the rays impinged.[83]

It had generally been assumed that electricity from a battery flowed from positive to negative, but these discoveries showed that the rays were equivalent to a flow of negative electricity from negative to positive. The question of whether this flow was in the form of rays or particles was not easy to decide, but eventually J. J. Thomson made a synthesis of the available evidence, added the results of his own research and settled the question to most people's satisfaction in favor of particles. Using magnetic and electrical deflections of the rays, in conjunction with work on ionic electric charges by his student, John Townsend, Thomson showed that the particles are very much lighter than the smallest known atom – that of hydrogen – and, later on, that they are identical with the beta particles from radioactive sources. In the course of his report he remarked:

> *I regard the atom as containing a large number of smaller bodies which I will call corpuscles; these corpuscles are equal to each other; the mass of the corpuscle is the mass of the negative ion* [cathode ray particle] *in a gas at low pressure.*

A few years later, Thomson, still pondering the nature of an atom that must now be thought of as composite, made what may have been the first attempt to construct a model that would account for the remarkable regularities of the periodic table. Lord Kelvin had suggested that the atom might consist of a sphere of positive electricity with electrons embedded in it, but Thomson tried to arrange the corpuscles, soon to be known as *electrons*, in concentric rings so that they would build up in a way that corresponded to the periodic chemical properties. Some years were to elapse before such ideas could be put into a viable form, and Thomson's corpuscles seem at first to have attracted very little attention.

(vi)
The Unstable Atom

In 1895 Wilhelm Conrad Roentgen showed that the impact of cathode rays on metal surfaces produces rays of a different kind. The new rays – X-rays – have remarkable penetrating power and can fog photographic plates through their usual protective layers of black paper. The early use of X-rays to produce photographs of the bones of living people created a sensation, and their discovery led indirectly to something even more momentous.

Antoine Henri Becquerel (1852-1908) was the son of Alexandre Edmond Becqerel (1820-1891), professor of physics at the *Conservatoire des Arts et Metiers* in Paris, and the grandson of Antoine Cesar Becquerel (1788-1878), an ingenieur-officier in Napoleon's army who became professor of physics at the *Musée d'Histoire Naturelle* in Paris. Antoine Henri succeeded to his father's position at the *Musée* in 1892, became professor at the Ecole Polytechnique in 1895, and discovered radioactivity in 1896.

While this genealogy suggests that destiny had taken a hand in preparing Henri for his momentous discovery, it cannot be denied that there was an element of fortuity in the process. As a result of the discovery of X-rays, Becquerel had become interested in the phosphorescence produced by the action of sunlight on various different substances, since the phosphorescence sometimes contained penetrative rays similar to X-rays. One substance that he tried was a double sulphate of uranium and potassium. He soon found that this substance did indeed emit penetrating radiation after exposure to sunlight. We may imagine his surprise, however, when he found that the radiation had nothing whatever to do with sunlight and continued unabated in a dark room. Further investigation showed that the active element was uranium.[84]

Like X-rays, ultraviolet rays, and electrical discharges, the radiations from uranium ionize the air and make it a much better conductor, allowing static electrical charges to dissipate. This property enabled Marie Curie (1867-1934) to measure the intensity of the radiations under controlled conditions and with her husband, Pierre, to discover other radioactive elements. The radiations are complex, their components penetrating matter to different degrees. In 1899, Ernest Rutherford found that some of the rays were stopped by an aluminum foil only a thousandth of an inch thick, a sheet of paper, or a few centimetres of air; others had a penetrating power comparable to that of X-rays, and could pass through several millimetres of aluminum. In his own words:

> *These experiments show that uranium radiation is complex, and that there are present at least two distinct types of radiation – one that is very readily absorbed, which will be termed for convenience the α radiation, and the other of a more penetrating character, which will be termed the β radiation.*

"The cause and origin of the radiations continuously emitted by uranium is a mystery," he went on to remark. Paul Villard soon discovered a third component of the radiation, which could pass through several centimeters of lead and was eventually named gamma rays. Textbooks often show the three kinds of radiation being separated by a magnetic field as if it were the easiest thing in the world, but as Rutherford remarks in his *Britannica* article; "A large amount of work... has been carried out to determine the nature of these radiations." It was not until 1903, seven years after the discovery of radioactivity, that Rutherford managed to show that alpha rays are deflected by electric and magnetic fields.

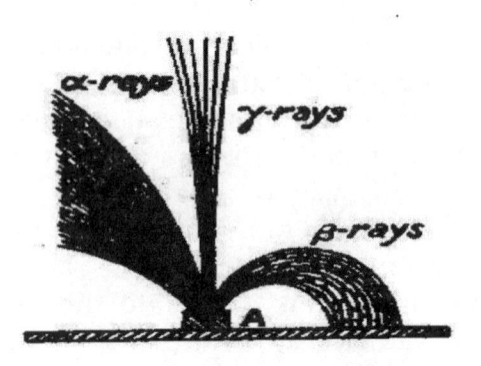

Figure 4:
Analysis of radium radiation by a magnet – reproduced from an early edition of Mellor's *Modern Inorganic Chemistry*

A beam of the radiation, obtained by placing lead sheets with small holes in front of the radioactive substance, falls on a fluorescent screen, producing a small spot of light. The effect of a magnetic field is to split the beam into three components – we now see three separate spots on the fluorescent screen, one in the position of the original spot and one on each side of it. It was concluded that uranium emits three kinds of radiation, which we still refer to as alpha, beta and gamma rays. Gamma rays correspond to the central spot on the fluorescent screen and are assumed to carry no electrical charge since they are not deflected by the magnetic field. Of the other two spots, the one closer to the original is associated with alpha rays and the one further off, on the other side, with beta rays. The interactions between the magnetic field and the alpha and beta rays indicate that the alpha rays consist of positively charged particles and the beta rays of negatively charged particles. *More precise work showed that beta rays are indistinguishable from cathode rays and that the alpha particles are the nuclei of helium atoms.* It would be hard to overstate the importance of these observations.

If the phenomena of the gaseous discharge tube were both startling and puzzling, it is easy to imagine that the discovery of radioactivity was downright stupefying. G. W. Mellor, the great chemical encyclopaedist, who was a young man at the time, quotes an anonymous contemporary:

> *The discovery that there are metals which, so to speak, are bleeding to death by the irrestrainable welling forth of strange aerial substances from their intimate parts was a novelty which held chemists spellbound with astonishment.*

The most general conclusions from all of this work on radioactivity and the conduction of electricity through gases were that atoms are composites of electrically charged particles and that they do indeed have an internal structure which can sometimes break. It is worth noting that the rate at which radioactive decay proceeds is not influenced by any known conditions of pressure, temperature or chemical combination – this is what makes radioactive dating plausible. Mellor quotes another anonymous writer, from 1907, as follows:

> *Radioactivity is the least manageable of natural processes. It will not be controlled. Nature keeps the management of*

this particular department in her own hands. Man views the phenomenon with hungry eyes, but his interference is barred out. He can only look on in wonder while it deploys its irresistible unknown forces.

By this time, the earliest versions of the Periodic Table were over forty years old, and while people looked with wonder at the latest developments in physics and chemistry, they realized that whatever model of atomic structure was proposed, it had to be capable of accounting, at least in principle, for the structure of the Periodic Table.

Chapter V
The Rutherford-Bohr Atom

(i)
Bohr Gets Involved

The first century of experimentation with electrical currents had ended about 1900 and up to that time there was no well-defined mental image of the inner mechanisms of electric circuits, so when people thought about electric currents it was generally with some vague idea of a fluid flowing along a pipe. In fact the laws governing simple electrical circuits are very similar to those governing systems of water pipes. The rate of flow is proportional to the applied pressure, and the amount that comes out of the end of a pipe is equal to the amount that went in at the beginning. An important development was the emergence in 1900 of the earliest ideas of solid state physics, in which Paul Drude suggested a primitive mechanism, elaborated by Hendrik A. Lorentz in 1905, for the conduction of electricity and heat through metals. Following the discoveries of the 1890's, the Drude-Lorentz theory proposed that the inner structure of a metal was built of spherical atoms that had lost some of the electrons that had originally belonged to them and that the "electrical fluid" consisted of a cloud of these unattached electrons that moved among the interstices between the spheres and could be propelled along a wire by the electromotive force of a battery.

The theory was modestly successful in explaining some of the features of the conduction of electricity and heat by metals but was unforthcoming with regard to several crucial issues, including the need felt by physicists to find some fundamental explanation for the nature of the energy radiated by hot metals. To put the matter another way, Planck's quantum still stuck in their throats and they wanted a physical theory that would do away with the need for it.

Niels Bohr (1885-1962), who was deeply interested in these problems and found the Drude-Lorentz theory unsatisfactory, suggested in his doctoral thesis at the University of Copenhagen that the reason why it didn't work very well was that it was based on an incorrect picture of the conditions inside the metal. He saw that something in addition to the existing mechanical and electromagnetic principles was needed in order to create a working model of atomic structure, but at this point he had no further suggestions to offer. After obtaining his Ph. D. in 1911, he moved to Cambridge, where J. J. Thomson had spent more than a decade trying to figure out the internal workings of the atom, and thence to Manchester to work with Rutherford shortly after the discovery of the nucleus.

To put the matter very briefly, Rutherford and his associates had found that thin metal sheets were almost perfectly transparent to alpha particles from a radioactive source, but very occasionally an alpha particle seemed to bounce off a heavy obstacle. Among high-speed particles on the atomic scale, the alpha particle is a positively charged monster, a helium nucleus with a mass[86] about 7,400 times that of an electron and 4 times that of a proton, so anything it bounces off must have at least comparable mass. Rutherford concluded that each atom in the metal sheets consists of a single, heavy, positively charged particle, which later became known as the nucleus, a cloud of electrons and a lot of empty space; but how the electrons were arranged around the nucleus and why they stayed in place was still a matter for conjecture.

It had been known for many years that the force between two electric charges is governed by an inverse square law of the same mathematical form as the law that governs gravitational attractions. Like Rutherford's picture of the atom, the solar system appears to consist mostly of empty space with a few relatively small objects circulating around a very massive central object, so it would have been quite natural to suppose that the atom resembles a miniature solar system, with the nucleus in the middle and the electrons orbiting around it. The problem about this model was that in order to keep moving in its curved orbit, an electron must always accelerate towards the nucleus.[87] As far as that is concerned, the electron is no different from a planet which is always accelerating towards the sun – otherwise it would just keep going in a straight line; the crucial difference for the atomic model is that the electron carries an electric charge and, according to 19th century electrodynamics, any accelerating electric charge acts like a radio transmitter, emitting energy in the form of electromagnetic waves. The electron would therefore lose its energy of motion and spiral into the nucleus. To put the matter very crudely – why is there all this empty space; why don't the tiny negative electrons simply stick to the massive positive nucleus?

(ii)
The Hydrogen Atom

Bohr's tentative views on atomic structure became known, and a colleague asked him whether he thought that Balmer's formula for the hydrogen spectrum might have anything to do with the problem. This was something that hadn't occurred to him before, and it set the wheels turning with so much energy that within a year he produced a substantially complete theory of the hydrogen atom. *Substantially complete*, it should be noted, is not the same as *definitive*, *final*, or *correct*.

The nucleus of the hydrogen atom was a single proton, with a single positive charge and a mass of one unit on the atomic scale. Tethered to it, in some way that was not at all clear, was a single electron with a charge equal and opposite to that of the proton, and only 1/1840 of its mass. This atom was the simplest possible object for Bohr's thinking, which incorporated three notions contrary to traditional physics. The first was that the atom can exist only in certain stable configurations, which Bohr called stationary states, in which the ordinary rules of mechanics apply but those of electromagnetism don't. These states were pictured as a set of possible orbits for the electron, their radii being calculated according to a quantum rule involving Planck's constant. The second was that energy in the form of electromagnetic radiation is emitted only when the atom falls from one stationary state to another – in other words, when the electron moves abruptly from a higher orbit to a lower. The third notion was that for each such transition, the radiated electromagnetic waves would be emitted as a single quantum with energy equal to the difference between the energy levels of the two states. This is where Bohr's ideas linked up with Planck's and Einstein's, and why the concept of energy levels became so important and familiar to students. The frequency of the radiated waves would be proportional to the energy difference between the two states in accordance with Planck's formula, $E = h\nu$. The absorption of radiant energy by an atom would reverse the process. A quantum would be absorbed if it contained exactly the right quantity of energy to raise the electron from one possible orbit to another.

These ideas incorporated an explanation of the line spectrum of hydrogen. Each separate line corresponds to a transition from one stationary state to another, and Bohr's rule for calculating the radii of the orbits leads to Balmer's formula, including the mysterious whole numbers, n and m. For an electron moving from one stationary state or orbit to another, m was the number of the orbit which it left and n the number of the orbit at which it arrived. Bohr's theory not only made possible the theoretical derivation of Balmer's empirical formula, which had puzzled physicists for thirty-seven years, it also enabled him to calculate the exact value of the Rydberg constant and to predict the existence of other sets of lines which would be found in the infra-red and ultra-violet regions of the hydrogen spectrum. Within a year, Theodore Lyman discovered the set corresponding to $n = 1$, meaning that the lines in this set were produced when an electron falls into the lowest orbit – number 1, or what Bohr called the ground state. Balmer's lines, according to these ideas, are produced by electrons falling from higher orbits into the second orbit. In addition to accounting for the Balmer Series, Bohr showed that his model of the hydrogen atom explained several perplexing phenomena involving the effects of electric and magnetic fields on the spectral lines. For a while everything looked rosy for it, in spite of the fact that its theoretical foundations seemed quite bizarre to the physicists of the time.

(iii)
Beyond Hydrogen

During the years 1913-1923 Bohr and other physicists put a great deal of energy into the development of his atomic model. Bohr's tendency to rely as much on intuition as on rigorous mathematics was rather alarming to some of his colleagues. One of the instruments of his intuition was the idea that when an electron passes from one orbit to another, the smaller the quantum leap, the more closely the result must correspond to what would be calculated using traditional electromagnetic laws. This idea, which enabled quantum relationships to be found by intelligent guesswork, was the origin of what became known as the Correspondence Principle, of which more in Chapter VII.

The brilliant initial success of Bohr's hydrogen atom could not be maintained, and attempts to apply his methods to the spectra of the helium atom were uniformly unsuccessful. Remember that the work we are discussing was basically all about line spectra observed under varying conditions, so the statement that the results for helium were discouraging means that the theoretically calculated frequencies did not agree with the experimentally observed ones. The situation went beyond mere discouragement; by 1923 Max Born had referred to this lack of agreement as a "catastrophe," and there was a general feeling that "some radical modification of the conventional quantum theory of atomic structure appears necessary."[88]

Bohr, however, was not one to be discouraged and from 1921 to 1923 he took his theory to a new stage in which he tried, with some success, to detail the atomic structures of all the elements. As with his original theory of the hydrogen atom, he used a mixture of whatever physical principles and data suited his purpose – X-ray spectra, chemical and physical properties, the structure of the Periodic Table, the calculated radii of the orbits and, of course, the Correspondence Principle. The feature of his model that gave the most promising connection to the Periodic Table was the idea of concentric shells of orbiting electrons. In the popular, simplified version each shell contains a number of electrons equal to the number of elements in the corresponding period of the table, but once we get past the third period the real story is not so simple. According to Bohr's application of the quantum theory, the numbers of electrons in successive shells should be 2, 8, 18, 32, 50, corresponding to the formula $2n^2$. For the first period, $n = 1$, so $2n^2 = 2$ and the shell contains just two elements, hydrogen and the inert gas helium. He therefore pictured the hydrogen atom as having an incomplete shell with one electron, and the helium atom a complete shell with two electrons. For the second period, $n = 2$, so $2n^2 = 8$ and the shell contains eight elements, from the alkali metal lithium to the next inert gas, neon. This suggested a second shell concentric with the first and having space for eight electrons. The chemistry of these elements is determined by the external form of the atom – in other words by the number of electrons in the outermost shell. The inertness of helium and neon is a manifestation of the fact that their outermost shells are complete and stable, while the hyperactivity of the alkali metals is due to the inviting presence of a single electron at the periphery. The following diagrams, adapted from the 1956 edition of *Modern College Physics*, by Harvey E. White, show the supposed electronic structures of the first four elements. Z is the atomic number, which equals the positive charge on the nucleus; n, the ordinal number of the orbit, is the principle quantum number and is the same n as the one in the Balmer's Series formula.

142 — *Rudolf Steiner and the Atom*

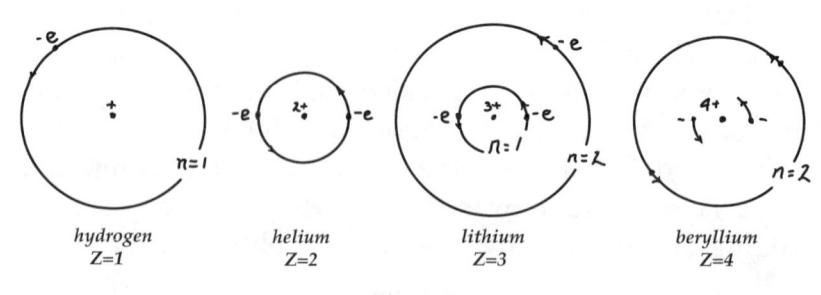

hydrogen	helium	lithium	beryllium
Z=1	Z=2	Z=3	Z=4

Figure 5

After neon comes sodium, an alkali metal quite similar to lithium in its physical and chemical properties and having one electron in its outermost shell. This is the start of another period of eight, ending with the inert gas argon and leading to the next alkali metal, potassium. One way of thinking about these elements is to say that we take a neon atom and add electrons one at a time until we have a new shell of eight electrons around it. These diagrams show the first three elements of the third shell.

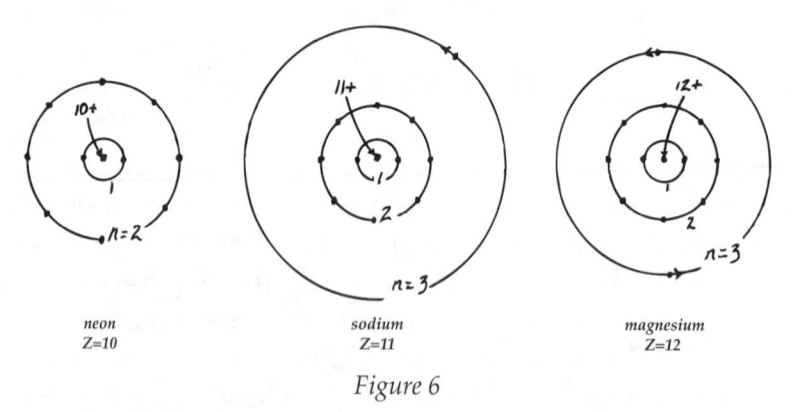

neon	sodium	magnesium
Z=10	Z=11	Z=12

Figure 6

Bohr's model suggested an explanation of the phenomena of valence that had been so important in assembling the Periodic Table. The outer shells of lithium, sodium and potassium each contain just one electron.

Beryllium, magnesium and calcium have two, while boron, aluminum and gallium have three, and carbon and silicon have four. Evidently for metals, even dubious ones, the valence is equal to the number of electrons in the outermost shell. Fluorine, chlorine and bromine, all univalent, have seven electrons in the outermost shell – in other words, it seems that one electron is missing from the number that would complete the shell. It was therefore concluded that the valence of a non-metal equals the number of electrons needed to complete the outermost shell. This line of thought linked up with the theory of chemical bonds put forward by Friedrich August Kekulé (1829-1896) in the 1860's and gave rise to the concepts of the ionic bond, in which electrons are donated by a metal atom and received by a non-metal atom, and the covalent bond, in which electrons are shared. This, again, was a nice picture but gave only a general idea of how bonds might be formed and a great deal more work was needed before the fact that many of the elements had more than one valence could be explained.

It is noteworthy that the problems of the earliest tables constructed by Newlands and Mendeleev began in the fourth period, which starts with potassium. Up to that point everything seemed to fit very well with the notion of electron shells, as the following table shows. As before, the atomic numbers are printed above the names of the elements, but now the numbers of electrons in successive shells are displayed below the names. Below silicon, for instance, in the middle of the third period, you will see 2/8/4, meaning that the atom has 2 electrons in the first shell, 8 in the second and 4 in the third. Potassium, 2/8/8/1, has 2 in the first shell, 8 in the second, 8 in the third and 1 in the fourth.

Group	I	II	III	IV	V	VI	VII	VIII
Period 1	1 Hydrogen 1							2 Helium 2
Period 2	3 Lithium 2/1	4 Beryllium 2/2	5 Boron 2/3	6 Carbon 2/4	7 Nitrogen 2/5	8 Oxygen 2/6	9 Fluorine 2/7	10 Neon 2/8
Period 3	11 Sodium 2/8/1	12 Magnes- ium 2/8/2	13 Alum- inum 2/8/3	14 Silicon 2/8/4	15 Phos- phorus 2/8/5	16 Sulphur 2/8/6	17 Chlorine 2/8/7	18 Argon 2/8/8
Period 4	19 Potass- ium 2/8/8/1	20 Calcium 2/8/8/2	....					

Table 2

At this point there is great potential for confusion (at which remark, the reader may give vent to a hollow laugh) as the *periods* designated by Mendeleev no longer coincide with the *shells* in Bohr's model. The third *period* of the table, which started with sodium, comes to an obvious end with the inert gas argon directly below the inert gas neon. The fourth begins with the alkali metal potassium directly below the alkali metal sodium. This suggests that the third shell is complete with eight electrons and that potassium resembles sodium in starting a new shell with a single electron orbiting outside a complete shell of eight. This impression is strengthened by the next element, calcium, which fits perfectly into Group II below magnesium. In Bohr's theoretical model, however, the third shell can actually accommodate 18 electrons (2 times 3^2) and for reasons which involve the fact that the energy of an orbiting electron depends on other factors in addition to the one quantum number that I have described, the fourth shell overlaps with the third. The result is that although the exterior electrons of potassium and calcium belong to the fourth shell, subsequent electrons go into the third shell one by one until it has its full complement of 18.

Group I		II		III		IV		V		VI		VII	
a	b	a	b	a	b	a	b	a	b	a	b	a	b
19 Potassium 2/8/8/1		20 Calcium 2/8/8/2		21 Scandium 2/8/9/2		22 Titanium 2/8/10/2		23 Vanadium 2/8/11/2		24 Chromium 2/8/13/1		25 Manganese 2/8/13/2	
	29 Copper 2/8/18/1		30 Zinc 2/8/18/2		31 Gallium 2/8/18/3		32 Germanium 2/8/18/4		33 Arsenic 2/8/18/5		34 Selenium 2/8/18/6		35 Bromine 2/8/18/7

VIII	0
26 Iron 2/8/14/2	
27 Cobalt 2/8/15/2	
28 Nickel 2/8/16/2	36 Krypton 2/8/18/8

Table 3

As the third shell fills up and the fourth is on hold with just one or two electrons, we have the sequence of metals mentioned in the previous chapter, now augmented by the ones discovered after the early versions of Mendeleev's table had been published. Scandium and titanium fit reasonably well in Groups III and IV, respectively, but vanadium, chromium, and manganese all have multiple valences and both metallic and non-metallic chemical properties and bear no particular resemblance to anything that has gone before. We have the impression that some of the electrons are not quite sure which shell they belong to, and that this is the cause of some of the ambiguous chemical behavior.

After manganese another complication ensues in the shape of iron and its siblings, cobalt and nickel, which Mendeleev had felt compelled to place in an entirely new group separate from the main body of the table. The shell of 18 is completed with copper (2/8/18/1) and we have a sequence of eight comparatively well behaved elements ending with the inert gas krypton, and echoing the sequence of the third period – sodium to argon. One of the features of Bohr's model is that within the larger shells there is always a privileged circle of 8, which helps to explain the presence of an inert gas at the end of each period and an alkali metal at the beginning of the next one. The subgroups that Mendeleev had to introduce in the fourth period are a consequence of the larger third shell.

As each shell accommodates more electrons than the previous one, the complications can only increase. The fourth shell has room for 32 electrons (2 times 4^2) and the consequences of this expansion show up in the Periodic Table in the form of the set of 15 rare earth elements. The periods and groups of Mendeleev's table correspond in less obvious ways to the numbers of electrons potentially occupying each shell, and the physical and chemical properties of some of the elements are less clearly appropriate to the families in which they are placed. Looking back historically, we note that Mendeleev had no idea what was going on inside the atom, and we marvel again at his pertinacity in ferreting out the relationships among the elements. Looking forward, we realize that while Bohr's model, improved with the help of other physicists, was successful in accounting for the overall shape of the table and many of its problematic features, there were still grave difficulties. The next step forward involved radical changes in the concept of the electron as an old-fashioned material particle and the simple notion of shells. How this came about is the main subject of Chapter VII.

The expansion of the sixth period of the Periodic Table had resulted from the discovery and inclusion of a large set of elements all very similar in their properties and known collectively as the rare earth elements. (See Ch. IV, Sec. (i).) These start with lanthanum, element No. 57 and end with lutecium, element No. 71 and lie between Group III and Group IV. Bohr proposed a way of grouping the electrons that plausibly accounted for the rare earth elements and suggested that the undiscovered element No. 72 would be found to have chemical and physical properties that placed it in Group IV below titanium and zirconium rather than with the rare earths. This conclusion was included in one of the lectures that Bohr gave in Göttingen on his latest theory in 1922, and within a few days, thanks to the work of two French chemists, he had to retract it. He did so with typical forthrightness:

> The only thing I know for certain as yet about my lectures in Göttingen is that several of the results I reported there are already wrong. The first point is the constitution of element 72 which, contrary to my expectations, has after all been shown by Urbain and Dauvillier to be a rare earth.[89]

Bohr was not entirely convinced, however. The two Frenchmen had looked in samples containing rare earth elements, but the Dane thought that the new element was more likely to show up in zirconium ores. Encouraged by Bohr, Georg von Hevesy and Dirk Coster used a new technique of X-ray analysis to find No. 72 among zirconium samples and showed that its properties placed it firmly in Group IV and not among the rare earths. Bohr had emulated his great predecessor, Mendeleev, in predicting the properties of a hitherto unknown element, now known as hafnium, after the Latin name for Copenhagen, and was able to include the announcement in his Nobel Prize acceptance lecture later in 1922.

Unfortunately, the hafnium triumph was only a temporary respite from the difficulties and miseries that Bohr's atomic model had generated, and it was even suggested that its successes had been attained more by luck than good judgement.

On the positive side it could still be maintained that the model actually worked very well in many respects and that it had made a good start on explaining the family relations of the elements in the periodic table. As far as elementary students and the interested general public were concerned, Bohr's atom had the great advantage that although it was mathematically complex, a simple picture of it was easy to visualize. This is what made it possible for the idea of an atom consisting of a central positive nucleus surrounded by a largely empty space in which electrons traveled in circular or elliptical orbits to penetrate the public consciousness, where it would become a fixed image and in the twenty-first century still appear in elementary physical science courses and popular science books. As far as the physicists were concerned, however, it was already out of date by the mid-1920's.

*

In its earliest days, the electron had seemed to be of merely theoretical importance, so much so that at an Annual Cavendish Laboratory Dinner someone proposed a toast; "The electron: may it never be of any use to anybody!" It wasn't long, however, before electrons seemed to be everywhere, and although very little was known about their habits, they caused so much excitement among physical scientists that in 1907 Henri Poincaré[90] was moved to observe, "The electron has conquered physics and many worship the new idol rather blindly."

This was true and it applied to the mental image of electrons flowing along a wire, which was only a first rough attempt at solid state theory. At first not many people outside the scientific community were aware of this theory, but the discovery of the nucleus and the development of the Bohr atom made modified versions of the Drude-Lorentz model quite plausible. What people like Niels Bohr and Arthur Eddington knew to be a conceptual model, subject to change, was apt to be taken by less perceptive citizens as a scientific fact, so there were soon thousands, perhaps even millions, of people who had simplified versions of electronic physics buzzing around in their minds. This was the situation when Steiner gave his last lectures on science.

Chapter VI
Late Words from Rudolf Steiner

(i)
A Science of Dead Matter

Now we return to the 1922 lecture cycle, which will lead us into the lecture to members of January 28th, 1923. (Referred to hereinafter as *1/28/23*.)

As we have already seen, Steiner continued to uphold the value of the modern scientific approach in his final public scientific cycle, given, as Owen Barfield tells us, "to an audience containing some professional scientists and others particularly interested in science, many of whom were members of the Anthroposophical Society." People who had attended his previous courses will not have been surprised by the following remark (which I have already quoted) but some of his subsequent statements must have caused a lot of heavy thought:

> ...*The scientific path taken by modern humanity [is] ... not erroneous but entirely proper... It bears within itself the seed of a new perception and a new spiritual activity of will.*

The lecture from which this remark is taken was given on Christmas Eve, 1922, in the first Goetheanum, exactly a week before its destruction by fire. Steiner, in what must seem to us to have been a superhuman effort, continued the course in his private studio, almost, as Barfield puts it, "as if nothing had happened," to which I would add, "Almost, maybe, but definitely not quite." The whole cycle is of extraordinary interest, but what I want to consider here for a moment is Steiner's emphasis on the perception that the physicists of his time were following the path in a direction that led only to a science of that which is dead or, perhaps, never had life in the first place.

The word "perhaps" in the last sentence is an acknowledgement of my inability to be absolutely precise about the distinction between being dead and never having been alive. A study of Steiner's *An Outline of Esoteric Science* suggests that even the most rigidly formed minerals still have some minute residue of consciousness, and he is careful to talk not about dead minerals, plants and animals but about "what is dead in them."

> *We have to become clear about what we actually do when, in our thinking, we cast inwardly experienced mechanics and physics into external space. That is what we are doing when we say: The nature of what is out there in space is of no concern to me; I observe only what can be measured and expressed in mechanical formulas, and I leave aside everything that is not mechanical. Where does this lead us? It leads us to the same process in knowledge that a human being goes through when he dies. When he dies, life goes out of him, the dead organism remains. When I begin to think mechanistically, life goes out of my knowledge. I then have a science of dead matter. We must be absolutely clear that we are setting up a science of dead matter so long as the mechanical and physical aspect is the sole object of our study of nature. You must be aware that you are focusing on what is dead. You must be able to say to yourself: The great thing about science is that it has*

> *tacitly resolved that, unlike the ancient alchemists who still saw in outer nature a remnant of life, it will observe what is dead in minerals, plants, and animals. Science will study only what is dead in them, because it utilizes only ideas and concepts suitable for what is dead. Therefore, our physics is dead by its nature.Science will stand on a solid basis only when it fully realizes that its mode of thinking can take hold only of the dead.*

I think it is also true to say that scientists not so tacitly resolved to recreate life, mentally at least, from the inanimate particles originally conceived in ancient Greece and given increasingly complex characteristics in the centuries that have elapsed since the Renaissance – in other words, by taking hold of that which is dead in order to understand, if not to reproduce, that which is living. Steiner does not criticize the scientists for concentrating on what is dead, but the process doesn't stop there and, as we shall shortly see, its further ramifications cause very great concern.

Later in the same lecture there is a comment about atomists:

> *In outer nature, those who proclaim atomism will always put you in the wrong. They even work themselves up to the very spiritual statement that when one speaks about matter in the sense of a modern physicist, matter is no longer material.... In this case they are right, and if we in our replies to them stop short at half-truths we shall never be equal to that which issues from them.*

Steiner had previously told us that the mathematical clarity with which we describe phenomena is a good thing, but that the sub-world of atoms that we then deduce is a sign not of insight but of mental inertia. If, however, we gather from this that the world of atoms and molecules is a mental construct with only the most tenuous physical existence and that in the light of further experience it will fall apart, we may be going a little too fast. As Steiner had emphasized in 1904, there is two-way traffic between the human soul and the natural world. We are changed by what we take in from nature and nature is changed not only by what we do to her but by what we think about her. Our conceptual world of atoms may unravel but not before it puts something into the world that will develop a life, and maybe a death, of its own.

(ii)

The Demonic Atom

Before quoting from Steiner's subsequent lecture to members, it may be helpful to recapitulate a little from the 1904 lecture.

> *When you build a house, you are inculcating human spirit into the raw material. If you construct a machine, you have laid the spirit that is part of you, into that machine; the actual machine does, of course, perish and become dust; not a trace of it will survive. But what you have done, what you have achieved, passes into the very atoms and does not vanish without a trace. Every atom bears a trace of your spirit and will carry this trace with it... Moreover, through your having changed the atom, through the fact that you have united the spirit in you with the mineral world, a permanent stamp has been made upon the general consciousness of mankind...*

In the 1/28/23 lecture, Steiner gives a vivid picture of the perils that await us when the atom is "electrified" and electricity becomes the basic explanatory medium for the natural world. This lecture seems to be unavailable in its entirety; the following comes from an excerpt printed in the News Sheet of the General Anthroposophical Society of

June 9th, 1940. The usual proviso, that this is from shorthand notes unrevised by the lecturer, applies here; Steiner was very well informed about contemporary physics, so it is unlikely that he spoke of transforming the atom "into an electron," but such lapses do not detract from the force and pungency of his message.

> The cultural ingredient that now permeates our whole external civilization began to rise to the surface at the turn of the 18th and 19th century... How long ago was that? — Less than 150 years ago, yet electricity is now a cultural ingredient. Indeed, it is far more than this! You see, when the men of my age were young fellows, not one of them dreamt of speaking of the atoms in the sphere of physics otherwise than of tiny, inelastic, or even elastic spheres colliding with one another, and so forth, and then they calculated the results of these collisions. At that time, no one would have dreamt of conceiving the atom without further ado in the way which we conceive of it today: namely, as an electron, as an entity consisting altogether of electricity.

> Human thought has spun itself altogether into electricity, and this occurred not so very long ago. Today we speak of the atoms as if they were small suns, centres around which electricity accumulates... Thus we suspect electricity everywhere, when we penetrate into the world's mechanism. This is where our civilization so closely connects itself with a definite manner of thinking. If people would not travel on electric tramcars they would not think that the atoms are full of electricity.

> If we now observe the connections that existed before the present age of electricity, we may say that they allowed the natural scientist of that time to experience the spiritual in Nature. Although scholastic realism had not completely vanished, electricity began to affect people's nerves, expelling

from them everything that tended towards the spiritual.

Things went still further. Even light, the honest light that surges through the world's spaces, was gradually defamed and brought into the ill repute of resembling electricity! You see, when we speak of electricity, we enter a sphere that presents an aspect to the imaginative vision different from that of the other spheres of Nature. So long as we remained within the light, within the world of sound, that is to say, in the spheres of optics and acoustics, it was not necessary to judge morally that which appeared in a stone, a plant, or an animal, either as colors in the sphere of light, or as sound in the world of tones; it was not necessary to judge these things morally, because we still possessed an echo of the reality of concepts and ideas. Electricity, however, drove out this echo. And if today we are, on the one hand, unable to discover a reality in the world of moral impulses, we are, on the other hand, even less able to discover a moral essence in that sphere which is now considered to be the most important constituent of Nature.

Today, if we were to ascribe a real power to moral impulses, if we were to say that they contain a force enabling them to become sensory reality in the same way in which a plant's seed becomes sensory reality, we would almost be looked upon as fools. And anyone who ascribed moral impulses to the forces of Nature would be looked upon as a complete fool! But if you have ever allowed an electric current to pass through your nervous system, so as to experience it consciously with a genuine power of vision, you will realize that electricity in Nature is not merely a current but that electricity in Nature is, at the same time, a moral element. When we enter the sphere of electricity, we penetrate simultaneously into a moral sphere. If you connect your knuckle at any point with a closed current, you will immediately feel that your

> *inner life extends to an inner sphere of your being, where the moral element comes to the surface, so that the electricity pertaining to the human being cannot be sought in any other sphere than that sphere which is also the source of the moral impulses. Those who can experience the whole extent of electricity, experience at the same time the moral element in Nature. Modern physicists... imagine the atom as something electric, and through the general state of consciousness of the present time, they are unaware that whenever they think of an atom as an electric entity, they must ascribe a moral impulse to this atom, indeed, to every atom. At the same time, they must raise it to the rank of a moral entity.... But I am not speaking correctly... for, in reality, when we transform an atom into an electron, we do not transform it into a moral, but into an **immoral** entity! Electricity contains, to be sure, moral impulses, impulses of Nature, but these impulses are immoral; they are instincts of evil, which must be overcome by the higher world.*
>
> *The greatest contrast to electricity is light. If we look upon light as electricity we confuse good and evil. We lose sight of the true conception of evil in the order of Nature, if we do not realize that through the electrification of the atoms we transform them into carriers of evil... When we think of them as atoms, in general, when we imagine matter in the form of atoms, we transform these atoms into carriers of death, as explained in my last lecture; but when we electrify matter, Nature is conceived as something evil. For electric atoms are little demons of Evil. This, however, does not tell us much. For it does not express the fact that the modern explanation of Nature set out along a path that really unites it with Evil.*

The path to which Steiner refers here is the nineteenth and twentieth century extension of the one that resulted from the great change in the perception of nature that took place in the late Middle Ages and the Renaissance. Nature was to be treated as an object of research, to be bullied into revealing its secrets, rather than as a living source of wonder, healing and delight. Scientists might still love the natural world, but only in their spare time. As professionals they could do whatever seemed profitable in the pursuit of scientific knowledge, no matter how much "constraint and vexation" it involved. This is not to romanticize former times; nature had certainly been exploited for other forms of profit, whether for the commercial extraction of metals or the more esoteric extraction of knowledge. But the purest and most moral studies of nature had always been of a gentler kind. Steiner describes this attitude in Lecture 13 of a cycle on the Mystery Centers[91] given in 1923.

> *There was one thing characteristic of these medieval scientists which to us is quite immaterial. It is thought today that anyone, good or evil, can work in a laboratory and make investigations. That does not matter a bit. He has the formulae and can make analyses and syntheses. Anyone can do it. But in those days when nature was regarded as the work of the divine, whether the divine in man or the divine in the great world of nature, the following demand was made: The man who investigates in this way must be filled with inner piety. He must be in a position to turn his soul and spirit to the divine spiritual element of the world. It was clearly understood that the inner secrets of the human being and nature would be revealed only to the scientist who approached his work reverently and prepared for his experiments as if preparing for a sacrificial offering. Scientists knew that if they had attained a state of inner goodness the questions which they asked would be gladly answered by divine spiritual beings.*

Many scientists continued to strive for inner goodness, but this striving became more and more divorced from their scientific work, even when this was done for the benefit of their fellow human beings. This meant that the operation of their intelligence was separated from their life of feeling, and however honorable their intentions, their science contained in itself no moral substance. As the divine disappeared from their perceptions of nature, it also disappeared from the content of their work, so that their object of study was now dead matter. This provided a wonderful stomping ground for those beings that Steiner called the "daemons of materialism." In 1/28/23 he tells us how people in the early Renaissance may have had some dim awareness of what was in the works.

> *Those strange people at the end of the Middle Ages, who were so much afraid of Agrippa von Nettesheim, Trithem of Sponheim, and others, so that they saw them walking about with Faust's malevolent poodle, expressed this very clumsily, but although their thoughts may have been wrong, their feelings were not altogether wrong. For, when we listen to a modern physicist blandly explaining that Nature consists of electrons, we are merely hearing him explain that Nature really consists of little demons of Evil! And if we acknowledge Nature in this form, we raise Evil to the rank of the ruling world-divinity.*

(iii)
Don't be an ostrich!

The contrast that Steiner makes between the electron as the carrier of evil and "the honest light that surges through the world's spaces" is given even greater point when we see that the electronic processes in the vacuum tube give out a kind of light that is quite different from sunlight or even the various kinds of lamplight that existed before the fluorescent lamp was invented. The old-fashioned incandescent electric lamp runs on electricity and the light that it produces is rightly described as artificial, but at least its hot filament produces a continuous spectrum of radiant energy quite similar to that of the sun. We can certainly save a lot of energy by using fluorescent bulbs, but we can't help wondering what the effects of constant immersion in what we might call electronic light may be. However, before we throw out all our fluorescent bulbs, we had better see what else Steiner has to say. Maybe we should keep them after all.

> *As modern men who do not proceed in accordance with old traditional ideas, but in accordance with reality, we would come across the fact that the electric element in Nature is endowed with morality in the same way in which moral impulses are endowed with life, with a life of Nature, so that, later on, they take on real shape, become a real world. In the same way in which the moral element one day acquires real shape in Nature, so the electric element once contained a moral reality. If we contemplate electricity today, we contemplate the images of a past moral reality that have turned into something evil.*
>
> *If Anthroposophy were to adopt a fanatic attitude, if Anthroposophy were ascetic, it would thunder against the modern civilization based on electricity. Of course, this would be nonsense, for only world-conceptions that do not reckon with reality can speak in that way. They may say: "Oh, this is Ahrimanic! Let us avoid it!" — But this can only be done in an abstract way. For the very people who thunder against Ahriman, and tell us to beware of him, go downstairs after their sectarian meeting and enter an electric tramcar! So that all their thundering against Ahriman, no matter how holy it may sound, is simply rubbish. We cannot shut our eyes to the fact that we must live with Ahriman. But we must live with him in the right way, that is to say, we must not allow him to have the upper hand....*

> *Ahriman and Lucifer have the greatest power over us if we do not know anything about them, so that they can handle us, without our being aware of it. The Ahrimanic electricity can therefore overwhelm civilized man only so long as he blandly and unconsciously electrifies the atoms and thinks that this is quite harmless. But in so doing, he does not realize that he is imagining Nature as a complex of little demons of Evil.*
>
> *When even the light is conceived of electrically, as has been done in a recent modern theory, then the qualities of Evil are attributed to the divinity of Good. It is really terrifying to see to what a great extent the modern contemplation of Nature has unawares become a 'demonology,' a worship of demons! We must be conscious of this, for the essential thing is consciousness: we live in the age of the consciousness soul.*

So, what is the consciousness soul and what do we do about Ahriman and Lucifer?

(iv)
The Struggle for Human Consciousness

This is supposed to be a book about Rudolf Steiner and atomic science, not a book about everything. As Jean Paul[92] remarked, "It is rather inconvenient that everything reminds one of everything else," but the fact is that if we are atomic scientists, anthroposophists, or both, we can't avoid the perception that all the phenomena of the cosmos are interrelated. This makes a problem for the author; it may be true that all roads lead to Rome, but we can only follow them one at a time and must bear in mind that they all also lead everywhere else. Like the traveler, the author has to follow one path at a time, but he also has to decide at what point it's time to go back and start on another one.

We may believe that everything started with the Big Bang, or that in the beginning God created the heavens and the earth, or that human beings were formed out of the warmth sphere of ancient Saturn, but in each case we must accept that the multiplicity of rocks, plants, animals, people, planets and stars that we experience stems from a common source. According to Steiner the common source is the creative work of the hierarchies and the members of the ranks of these high spiritual beings have not always agreed with one another about how the process of evolution should proceed. Our progress towards spiritual freedom has further been hindered by the Luciferic spirits who would prefer us to be *automatically* good and spiritual and by Ahriman and his minions, who wish to destroy our perception of the spirit and confine us to a purely physical existence.

The interrelatedness that is present from the beginning is reinforced by the interdependence that arises from occupying the same space and sharing the same resources. The chipmunk who is displaced when a tract of open countryside is needed for the construction of a mall, doesn't know that he is sacrificing his livelihood for the sake of human commerce. It is probable that a considerable proportion of those involved in the project and those who make use of the result have no compelling religious or philosophical beliefs and no particular interest in the origin of the universe or their kinship with their co-inhabitants, including the chipmunk. This lack of consciousness, with which we are all, to some extent, afflicted, extends to matters that go far beyond the construction of malls. We may be honest, moral people who deplore the rape of the countryside, the squandering of the earth's resources and the exploitation of our fellow men and women, but we may still be quite unaware of the spiritual realities behind these activities. The chipmunk is blessedly free from the obligations of consciousness but the human being who is thus unconscious is open to the insidious influence of the two powerful spiritual beings named by Steiner. "Ahriman and Lucifer have the greatest power over us if we do not know anything about them, so that they can handle us, without our being aware of it."

*

In his lectures on the Karma of the Anthroposophical Society, Steiner spoke of the great battle between the powers under the leadership of the Archangel Michael, who wish for the free and healthy evolution of the human race, and those led by Ahriman. Michael is one of seven Archangels who guide and direct the fundamental tendencies of successive ages in relation to man, each one occupying the leading position for a period of between three and four hundred years. Michael was involved in – in fact, we might say "supervised" – the changing relationship of the divine intelligence to the human being from the very beginning and at the crucial time when the earliest stirrings of Greek philosophy took place he became the leading Archangel. His reign began in the pre-Socratic period, lasted through the age of Socrates, Plato, Aristotle and Alexander the Great and ended soon after the death of Alexander in 323 B.C. Steiner describes Aristotle and Alexander as Michaelic figures and they continued to work from the spiritual world for the healthy evolution of humanity. Michael's influence did not end at that point, but he had to work from a more remote region, and his relationship to the administration of the cosmic intelligence changed.

While Michael was preparing for his next period of rulership, the Ahrimanic spirits of materialism from the lower regions of the earth were doing their utmost "to prevent Michael's dominion from prevailing on Earth. And at that time the Ahrimanic spirits whispered to those who would lend their ear: The Cosmic Intelligence has fallen away from Michael and is here, on the Earth: we will not allow Michael to resume his rulership over the Intelligence...

Towards the end of the 19th century Michael himself would once again assume dominion upon the earth, but this new Michael Age must be different from the others. For what Michael had administered through many aeons had now fallen away from him. But he was to find it again when at the end of the seventies of the 19th century he would begin his new earthly rule. He would find it again at a time when an Intelligence intensely exposed to the Ahrimanic forces and bereft of spirituality had taken root among men. For while the Intelligence was descending from the cosmos to the earth, the aspirations of the Ahrimanic powers grew ever greater, striving to wrest the Cosmic Intelligence from Michael.

Such was the crisis from the beginning of the 15th century until our day, which expresses itself as the battle of Ahriman and Michael. For Ahriman is using all his power to challenge Michael's dominion over the Intelligence that has now become earthly. And Michael, with all the impulses that are his, though his dominion over the Intelligence has fallen from him, is striving to take hold of it again on earth at the beginning of his new earthly rule... So Michael finds himself obliged to defend against Ahriman what he had ruled through the aeons of time for the benefit of humankind. Mankind stands in the midst of this battle; and among other things, to be an anthroposophist is to understand this battle to a certain extent at least.

Meanwhile, Lucifer is still around, and he and Ahriman seem to have made a pact. As Steiner explains, the two have managed what we used to call a pincer movement in the Second World War. Steiner describes their activities in a lecture called *The Work of the Angels in Man's Astral Body,* given in Zurich in 1918.

Through the Angels, the Spirits of Form are shaping pictures in our astral body, pictures which contain forces for the evolution of mankind – the impulse to brotherhood, to the perception of the hidden *divinity* in our fellow human beings, and the possibility of *reaching the Spirit through thinking,* to cross the abyss and through thinking to experience the reality of the Spirit. Unless we become conscious of this work, however, things will go radically wrong.

> We are "*heading towards the time when purely through the Spiritual (Consciousness) Soul, purely through our conscious thinking, we must reach the point of actually perceiving what the Angels are doing to prepare the future of humanity... But the progress of the human race towards freedom has already gone so far that it depends upon us whether we will sleep through this event or face it with wide awake consciousness...*"

> *If people study Spiritual Science more and more thoroughly, if they assimilate its concepts and ideas, their consciousness will become so alert that instead of sleeping through certain events, they will be fully aware of them. Then a threefold truth will be revealed to mankind by the Angels."Firstly, we shall learn to understand the deeper side of human nature, to see spiritually what out fellow human beings really are. That is the one point— and that is what will particularly affect the social life.*

> *Secondly: The Angels will reveal to us that the only true Christianity is the Christianity which makes possible absolute freedom in the religious life.*

> *And thirdly: Unquestionable insight into the spiritual nature of the world.*

If we follow the dictates of our proper nature, we can not very well fail to perceive what the Angels are unfolding in our astral bodies; but the aim of the Luciferic beings is to tear us away from insight into the work of the Angels.... Lucifer wants us to be good and spiritual, but with automatic goodness and spirituality. He wants to lead us automatically to clairvoyance, to remove from us the possibility of evildoing, so that we act out of the spirit, but as reflections, as automata, without free will... so that we shall sleep through the impending revelation.

But the Ahrimanic beings too are working to obscure this revelation by destroying the consciousness of our own spirituality and convincing us that we are nothing but completely developed animals. Ahriman is the promoter of materialistic Darwinism, total technology and the refusal to acknowledge the validity of anything except the external life of the senses. The Ahrimanic beings are endeavouring to darken in man the consciousness that he is an image of God.

From this mention of the streams which run counter to the normal, God-willed evolution of the human race it can be gathered how we must conduct our lives, lest the impending revelation find us asleep. If we are not alert, the evolution of the Earth and the human race will be in grave danger.

The angels are working to achieve something that can be fulfilled only in earthly humanity, and only if we are awake to what is happening. If we sleep through the spiritual processes and events of our time, the angels will have to fulfill their aims without our knowledge. This was the great danger for the age of the Spiritual Soul about which Steiner spoke in 1918, and it might still happen if, *before the end of the century*, people were to refuse to turn to the spiritual life. He explains what the results of this process would be, and now that the end of the century has come and gone, we can get some idea of the validity of his insights.

First, certain instincts connected with the mystery of birth and conception and with sexual life as a whole would arise in a pernicious form instead of wholesomely, in clear waking consciousness. These instincts would pass over into the social life and would prevent us from unfolding brotherhood in any form whatever on the earth.

The second aspect is that while everything connected with medicine will make a great advance in the materialistic sense, people will acquire instinctive insights into the medicinal properties of certain substances and treatments—and thereby do terrible harm. But the harm will be called useful. People will actually *like* things that make the human being—in a certain way —unhealthy. It will then be possible either to bring about or not to bring about illnesses, entirely as suits their egotistical purposes.

The third result – the one that is most obviously germane to the discussion of Steiner and the atom – will be the discovery of forces which, by bringing certain vibrations into accord, will enable us to unleash tremendous mechanical forces in the world. Spiritual guidance and control of mechanical principles will lead technical science into a wasteland which human egotism will find useful and beneficial.

There is no need to labor this point. People may legitimately differ over matters of detail, but here we can upend a conventional sentiment. The devils don't care that much about the details; what they're interested in is the overall effect. We may mourn the death of a solitary chipmunk who embodies the wisdom of aeons of patient work on the part of the hierarchies, but whatever satisfaction

Ahriman, Lucifer and their cohorts get out of genocides, mass extinctions and the fouling and destruction of our planet is small beer compared to their joy at the prospect of enslaving the human race and taking the divine intelligence for themselves. And all this comes about because, as Steiner said in 1904, we have yet to learn and practice selflessness.

As Steiner says, this knowledge of human evolution can be truly understood only through a spiritually informed view of life. An unspiritual conception of life would give no understanding and would regard these developments as evidence of superhuman progress, of freedom from convention. In a certain respect, ugliness would be beauty and beauty, ugliness. Nothing of this would be perceived because it would all be regarded as natural necessity.

> *If we understand how Spiritual Science affects our whole attitude of mind, there can arise the **earnestness** required for receiving such truths, leading to the acknowledgment of definite responsibilities in life…*

This means that we must learn to know this Ahriman, who strives to take the divine intelligence for his own purposes and tries to snare us by encouraging us to do the same. To return to the Karma Lectures:

> *Ahriman stands before us as a cosmic Being of the highest imaginable Intelligence, one who has already taken the Intelligence entirely into the individual, personal element. If ever we let ourselves in for a discussion with Ahriman, we should inevitably be shattered by the logical conclusiveness, the magnificent certainty of aim with which he manipulates his arguments. In Ahriman's opinion, the really decisive question is this: Will cleverness or stupidity prevail? And Ahriman calls stupidity everything that does not contain Intelligence within it in full personal individuality.*

> Michael, however, is not in the least concerned with the personal quality of Intelligence. We human beings are always tempted to make our Intelligence personal as Ahriman has done. But Michael wills only to administer the Cosmic Intelligence and not to make it personally his own. And now that people have the Intelligence, it should again be administered by Michael as something belonging to all mankind — as the common and universal Intelligence that benefits all of us alike.

This last comment may seem confusing. We have learnt from Steiner how the universal intelligence came down into the hearts and minds of individual people, and how thinking became an individual matter; now he is speaking about "the common and universal intelligence that benefits all of us alike." The point is that in the distant past the divine intelligence thought in us and, for good or ill, there was nothing we could do about it. The good spirits have given us the freedom to use our thinking in any way that we wish – selflessly as individuals for the good of mankind, selfishly for our own private purposes or casually and carelessly as the whim takes us. The more selfish and the more casual we are, the more likely we are to hand our freedom over to Ahriman.

Knowledge of the spirit enables us to take Ahriman's weapons and turn them against him and to avoid the Luciferic temptation to wallow in a warm bath of pseudo-spirituality and dwell only on what makes us feel happy. We embrace the battle between beauty and ugliness, we seek the spirit that remains in the natural world and we are aware that, in Steiner's words:

> Behind the scenes of existence is raging the battle of Michael against all that is of Ahriman. And this is among the tasks of the anthroposophist... to feel that the cosmos is as it were

in the very midst of the battle. But Michael insists that his dominion shall prevail at any cost. Michael is a Spirit filled with strength, and he can only make use of brave people who are full of inner courage.

This would be a fine note on which to end a chapter, but I have to admit that when someone tells me that I have to be brave, the first thing that happens is that I feel scared. I don't feel like the sort of person on whom the future of the human race ought to depend. So here are a couple of encouraging thoughts, of which the first is that one of the ways in which Michael operates is to give people courage. Steiner talks about the benefits of studying spiritual science and it is especially heartening to read about the Archangel Michael, which you can do in *The Archangel Michael – His Mission and Ours*, edited by Christopher Bamford, in the Karma Lectures, in the collection of Michaelmas Lectures, in the *Letters to Members*, in the *Four Seasons and the Archangels* and in many other places. It strikes me that Michael is not the sort of Archangel who embraces lost causes, but one who takes the initiative and never lets go. Steiner says that he is determined to win whatever the cost, and the cost may be very great. Whatever catastrophes may lie in the future, the most important thing is that we keep our souls intact. I don't say pure – we are all tainted in one way or another – but in some part still devoted to what is good, true and unselfish. And for this we have help from the great Source of all sources, of whom Michael is the representative.

(v)
So what about the electron?
How does the thinking of the atomists and quantum physicists look in the light of Steiner's insights into the purposes of Ahriman, Lucifer and Michael? And how, in

that light, do we understand his remarks about electrons in 1923?

I've already quoted Poincaré's assessment of the situation very early in the 20th century: "The electron has conquered physics and many worship the new idol rather blindly." At that time only two fundamental particles were known, the electron and the proton, and the only fundamental forces were the gravitational and the electrical. There seemed to be no way of incorporating gravity into atomic physics, a situation that is still with us, so the atom consisted of protons (positive) and electrons (negative) held together by electrical forces. The nucleus was an altogether mysterious entity, carrying almost the whole mass of the atom but occupying only a tiny fraction of its volume, and apparently consisting of a number of protons and a smaller number of electrons, united in some fashion that no one could explain. Except in the case of the few radioactive elements, it appeared to sit placidly in the middle, taking no part at all in whatever adventures the atom might get up to. All observed electrical and chemical phenomena were explained – if explanations were available – in terms of what were then known as the "extra-nuclear" electrons, the ones outside the nucleus. As the years went by this was extended to the operation of the nervous system and the biochemical functioning of living organisms, so it wasn't only physics that the electron conquered. Anyone with any interest in science was likely to be bombarded, metaphorically speaking, with electrons. As far as popular science (with a few exceptions) and high school and college physics[93] were concerned, the situation didn't change much even when the quantum mechanics of the later nineteen-twenties and early thirties began to replace the old-fashioned mixture of quantum and classical physics and multitudes of new particles were discovered.

The neutron, discovered by Chadwick in 1932 and having zero electric charge and almost the same mass as the proton, seemed to be a thoroughly respectable particle which we could welcome because it made it just a little easier to understand the constitution of the nucleus. But the so-called "cosmic ray" particles, first suspected as a result of radiation measured at the top of the Eiffel Tower in 1910, later detected by exposing photographic plates on wild mountainsides and eventually created in high-energy labs, seemed to have no *raison d'être*. They had peculiar masses, in between those of the proton and the electron, and were not constituents of the atom as we understood it. As undergraduate students in the early 1950's, we rather resented them, called them "strange" or "funny" particles and would have liked to send them back to the mountainside like unwanted children. Somehow we had missed the fact that the great Japanese physicist Hideki Yukawa had already (1949) been awarded the Nobel Prize for physics for his prediction, in 1935, of the existence of the π-meson, now known as the pion, a "funny" particle that plays an essential part in holding the nucleus together. The Bohr atom that we had known and loved in High School seemed so right, proper, and beautiful that we had great difficulty in facing the fact that it was thirty years out of date, and another year or two had to pass before we learned better. In other words, what Poincaré had said was true; the electron had got into our brains and stuck there in its pre-Heisenberg form, a great example of the "dead concept" that Steiner warned teachers about in his education lectures.[94] And if we take his *1/28/23* lecture seriously, we can't help asking what it carried with it.

> *We suspect electricity everywhere, when we penetrate into the world's mechanism... When we electrify matter, Nature is conceived as something evil. For electric atoms are little*

demons of Evil. And if we acknowledge Nature in this form, we raise Evil to the rank of the ruling world-divinity.

The electronic atom, conceived and developed by highly accomplished, well-meaning physicists from the late 1890's to 1920, carried its Ahrimanic baggage into the minds of eager students and an unsuspecting public throughout the twentieth century. The fact that this oversimplified picture had a constrictive effect on people's thinking did not prevent physicists and engineers from producing sensational results, and it has to be recognized that, up to a point, an incorrect or incomplete theory may operate quite successfully in the physical world. The electrical engineering of the nineteenth century and much of the twentieth (including the design and manufacture of "tramcars"!) was based on rules that had been discovered in happy ignorance of the electron. The electronic theory of conductors and semiconductors was still quite crude when radio tubes and transistors were developed; experience and rules of thumb were perhaps more important than any calculations that could be based on the theory – but the radios worked very well. The nuclear weapons of 1945 were created with very little knowledge of the structure of the nucleus. The relation between theory and practice is extraordinarily complex. Matter may or may not be material, but it does have an awkward habit either of correcting people's assumptions or of playing them along until some denouement occurs which may be anything from comical to disastrous.

Modern physicists may have taken a wrong path, and they may be unconsciously furthering Ahriman's purpose, but there is still something good and human in their impulse to understand. Elements of this understanding pass into the world at large, generating technological and engineering projects, and carrying the Ahrimanic virus, which then infects people who don't understand what's going on. In the first half of the twentieth century "science" was a magic word. For many people, anything that the scientists said and backed up with nice pictures of atoms and molecules came into the category of "progress" and could be taken as true and good for society. When the Soviet Union beat the United States to the punch with the successful launch of a sputnik in 1957, the desire to regain the lead produced a great surge of enthusiasm for science and technology that has gradually dissipated in the face of huge expense, collateral damage and growing disillusionment. Now there may be more people who are awake and sceptical, but the enthusiasm has tended to be replaced by apathy, a condition that must be very pleasing to Ahriman.

In saying this, I am not consigning all technology and engineering to the devil. This is the nature of life in the modern world, and "thundering against Ahriman" doesn't do any good, no matter how vividly we picture Ahrimanic imps riding on virtual electrons. Being infected by the virus doesn't automatically make you a bad person; but if you are unconscious of what's happening, you keep getting sicker and sicker until you are completely under Ahriman's thumb. What starts out with the best intentions always has the potential to turn to evil. When Michael Faraday and Joseph Henry did their pioneering work on electromagnetism in the 1830's they had no idea that their discoveries were part of a stream that would eventually lead to the modern vehicles of mass communication that have given Ahriman all he needs for the purposes of propaganda and brainwashing. It may be a shame that political and commercial interests have gained control of nearly all the news media, but this is nothing compared with the influence of the picture of the human being as a fairly advanced and utterly selfish animal, devoted to the exploitation of sex and violence, that is continually washing into the human unconscious. This is not the place to enumerate the evils that result from the kind of science we have produced in the past few centuries and the uses to which we put it, but only to remember that the final purpose of spiritual activity is the salvation of souls and spirits. No one wishes to be drowned, incinerated, starved or poisoned in a natural or man-made catastrophe, but the loss of one's soul is a far more serious matter and that is exactly what Ahriman is aiming at. *"For what shall it profit a man, if he shall gain the whole world and lose his own soul?"*[95]

> *Ahriman and Lucifer have the greatest power over us if we do not know anything about them, so that they can handle us, without our being aware of it.*

When the divine powers allowed the cosmic intelligence to descend into our hearts and minds, they undoubtedly intended us to use it. One use for it is to recognize when discussion and persuasion are not only hopeless but also dangerous. While specifically advising us not to shun the methods and artifacts of modern science and technology, Steiner also warned us of the perils of getting into conversation with Ahriman:

> *If ever we let ourselves in for a discussion with Ahriman, we should inevitably be shattered by the logical conclusiveness, the magnificent certainty of aim with which he manipulates his arguments.*

Ahriman has many representatives on earth through whom he speaks very persuasively, sometimes appealing to our more selfish instincts and sometimes giving a very convincing appearance of taking the moral high ground; but if our heads and hearts are working according to divine purpose, we can recognize the source and find the courage to say, "Get thee behind me, Satan!"

*

In the light of Steiner's remarks about the electron, it is possible to regard the emergence of new forms of quantum theory in the 1920's as a healthy corrective. Up to that time, atomic theories had always been mechanical and deterministic. Events proceeded independently of the observer and continued on their predetermined paths whether or not we took any interest in them. Sometimes we thought about them, but thought was not in them. As quantum physics developed, thought became part of the fabric of the theory; the natural world appeared to some of the greatest physicists to be inextricably entangled with human activity. This may look like a return to some kind of philosophical realism, but we must remember that the thought that the old realists found in the natural world was divine, not human. It's true that human thought came from a divine source, but this is not the kind of argument that a quantum physicist would be likely to make. Furthermore, the physical world as implicitly defined by the quantum physicist is not the natural world of everyone else, and it may never be. It may be hoped that exploration of its strange, abstract territory will eventually enable us to reconstruct the comforting world of everyday experience, with ourselves as the center of it only because we did the thinking; but at this point we are very apt to feel like accidents which the universe has absentmindedly coughed up and whose only possible purpose is to provide some applause for its virtuosity. At this point it is a good idea to ponder Steiner's remark that the world is a riddle and the human being is the answer.

Chapter VII
The Atom After Steiner

(i)
Waves and Particles

I have sketched some of the developments in early twentieth century physics which Steiner was probably aware of and which were related to his concerns about the destiny of the human race. His remark, already quoted, that physicists "even work themselves up to the very spiritual statement that when one speaks about matter in the sense of a modern physicist, matter is no longer material.... In this case they are right...." shows that he knew that something radical was in the works. By the time of his death in March, 1925, it had not only become impossible for physicists to hold on to the old view of the electron and the proton as incredibly small bits of matter that carried electric charges and obeyed the same dynamical laws as large bits of electrified matter; it had begun to look as if the whole idea of a visualizable conceptual model of the internal workings of the atom would have to be abandoned. The Bohr atom, with its central positive nucleus surrounded by a nest of orbiting electrons, was a pretty picture, but it left far too much unaccounted for and there were several other developments that threatened to throw atomic physics into chaos.

The Atom After Steiner — 177

It is no exaggeration to say that it would take a huge volume and a team of writers to deal adequately with the adventures of atomic physicists in the period from the first version of the Bohr atom in 1913 to the time in the early 1930's when fully worked out versions of quantum physics were in use, and a tome of even mightier proportions to cover subsequent excursions. What I would like to do is to convey the strangeness of the quantum world that developed in this period and to raise, although not to answer, the question of the reality of that world.

*

To recapitulate a little: Max Planck invented the quantum as a way of generating an equation that matched the experimental observations of the energy radiated from a black body. Previously it had been thought that radiant energy traveled as continuous waves, but according to Planck's formulation, energy *left* the hot body as quanta and according to Einstein's theory of the photoelectric effect, energy *arrived* at the sensitive material in the form of quanta. It might therefore seem natural to suppose that it had traveled as quanta, that the quantum was a physical reality and that radiant energy, including light, always travels as quanta; but these ideas were received with a great deal of scepticism.

By 1915, Robert Millikan, who set out to disprove Einstein's theory finally had to admit that Einstein's equation was correct "in spite of its unreasonableness," but he still couldn't swallow the concept behind it. Let me remind you that when scientists' thinking carries them into apparently arbitrary or fantastical imaginings, it is often not simply through the momentum that Steiner mentioned, but "under the compulsion of observation," to repeat Max Born's phrase. This was how we got the quantum in the first place. Planck didn't like it any more than Millikan did, but finally he had to acknowledge that it was the only idea that worked. This doesn't prove that it was true, where "true" means "corresponding to something actually present in the physical world," but that it was something that physicists could work with and might have verifiable consequences. One of the consequences, as shown by Einstein, was that these quanta, which later became known as photons, must be able to act as concentrated points of energy – in other words, very much like particles.

A particle is the exact opposite of a wave; a wave, whether its medium is water, air, or an electromagnetic field, spreads out in all directions from its point of origin until it happens to encounter an obstacle; it potentially occupies the whole ocean, the whole atmosphere or the whole of space. It is not possible that the energy of the entire wave should act at a single point, but that is exactly what appears to happen with the photoelectric effect and in other circumstances, too. Sometimes a photon collides with an electron, which is something inconceivable for a wave to do, and we get the Compton effect, first described in 1923. So we have the paradoxical situation in which a wave spreads throughout a huge volume and may be detected at any point in that volume, and yet under certain circumstances, the energy of the entire wave acts at a single point as though it were concentrated in a particle. To put it another way, a beam of light[96] sometimes behaves like a stream of particles (photons) and sometimes like a succession of waves. This was very confusing and was one of the signs that the mental images we have of waves and particles might be inappropriate, inapplicable or even irrelevant when we are dealing with subatomic phenomena. If the concepts appropriate to modern physics represent reality, it follows that reality is much weirder than we had expected.

To compound the weirdness, Count Louis de Broglie,[97] in 1924, proposed that if light shows a duality between wave and particle behavior, we may expect streams of particles such as electrons to do the same thing. We used to be quite sure that light traveled as a wave motion, but now it sometimes behaves like a stream of particles. For twenty-five years we were sure that cathode and beta rays were streams of particles – electrons – but now de Broglie was suggesting that they have wave characteristics as well. Not content with merely floating this idea, de Broglie actually predicted the frequencies of the waves and experiments were soon designed to test his hypothesis, using the phenomenon of diffraction.

One of the pillars of the wave theory of light was the fact, already known at least as far back as the seventeenth century, that the shadows cast by light are not quite geometrically exact; as the light passes an obstacle it spreads a little into the shadow and the shadow spreads a little into the light, forming what is known as a diffraction pattern. This phenomenon takes place all the time in nature, but on such a tiny scale that it is rarely noticed. If the light is constrained to pass through a very small hole in an opaque screen, the effect is magnified and can easily be photographed.

According to old-fashioned ideas, particles such as electrons traveling through such a hole ought to keep going in straight lines, but if de Broglie's idea is correct, a beam of electrons passing through a small opening ought to spread out and produce a diffraction pattern analogous to that of light, from which the frequency of the associated wave could be calculated. This kind of set up wasn't practical for electrons, but in 1926, Clinton Davisson and Lester Germer in the USA and George Thomson in England obtained electron diffraction patterns by different methods to confirm de Broglie's predictions.[98]

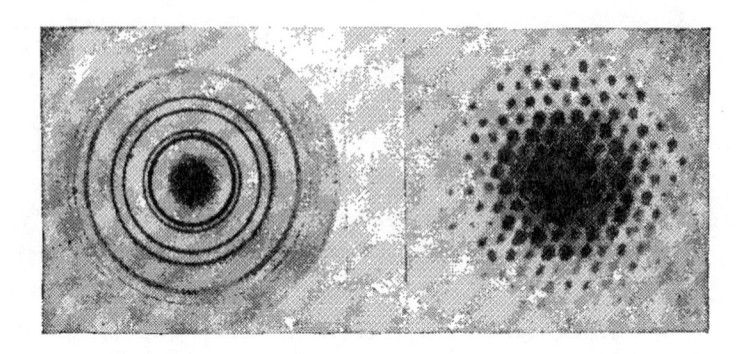

Figure 7:

Diffraction patterns from electron beams passing through silver foil and thin mica crystals

George Thomson was the son of "JJ", who had identified the electron in 1897. One of the pleasant ironies of history is that in 1906 Thomson *père* was awarded the Nobel Prize for showing that electrons behave like particles, and in 1937 Thomson *fils*, along with Davisson, was awarded the prize for showing that they behave like waves.

This was one of several events that made some of the physicists realize that old-fashioned mental images of sub-atomic processes were no longer productive or even possible. At one time we thought that the physical world could be described in terms of two different and separate kinds of thing – electrically charged material particles and electromagnetic waves. When the charged particles moved in certain ways, waves were generated, but a particle was still a particle and a wave was still a wave. This division was a convenient way of thinking, but in the period from 1905 to 1924 it was gradually seen to be an illusion. Whether light behaved like waves or like particles, and whether a stream of electrons behaved like particles or like waves seemed to depend on the circumstances or, to put the matter more disturbingly, on what the observer was doing. So it became very difficult to maintain any kind of mental image or model of what was going on in the atomic and sub-atomic world.

The grand old method of creating a conceptual model – like Newton's atoms "fleeing from each other" – and mathematically investigating its properties had been extremely successful. It had provided wonderful insights, from the early kinetic theory to the latest theories of thermal radiation, electrical conduction and the Bohr atom, but now it seemed to have shot its bolt. One of the problems of the early nineteen-twenties was that the physicists were trying to deal with a whole range of new or fairly new phenomena that involved quantum theory or the still very novel Einsteinian relativity theory with mental images, like the orbiting electron, that were inextricably connected with classical Newtonian mechanics and nineteenth century electrodynamics. To put it more briefly, physics was a mess concocted of jarring concepts and methods arbitrarily chosen to fit particular circumstances.

(ii)
Knabenphysik

When I was a student at Cambridge in the 1950's, the general opinion was that a physicist who had not made a significant breakthrough by the age of thirty was dead in the water and, in any case, you couldn't expect much from anyone over that age. Newton, who did all his best work in his twenties, was cited as an example; but the main evidence was the group of youthful geniuses who were active in the 1920's, including Enrico Fermi (1901-1954) and the three young men responsible for the nickname *Knabenphysik* ("boy-physics"), applied to the "new" quantum physics (as opposed to the "old" quantum physics of Planck and Einstein); Wolfgang Pauli (1900-1958), Werner Heisenberg

(1901-1976) and Paul Dirac (1902-1984). Einstein had already set the bar pretty high by publishing three great papers at the age of twenty-six and Bohr had produced his hydrogen model at twenty-seven; and it was conveniently forgotten that Max Planck was over forty when he invented the quantum and that J. J. Thomson and Ernest Rutherford were both forty when they made their most well-known discoveries. Anyone who happened to remember this was told "Well, things are different in the twentieth century." Somehow it was never mentioned that Niels Bohr and Max Born played leading parts in the development of the "new" quantum theory when they were in their forties, and Erwin Schrödinger (1887-1961) was thirty-eight when he produced his celebrated wave equation. We were not to know that Richard Feynman (1918-1988) would continue to produce highly original, fundamental work well into his fifties. What attracted us to Pauli, Dirac, and Heisenberg was not only their youth, but also the "otherness" of their conceptions. It was a little like one's first encounter with anthroposophy.

(iii)
"Thou shalt make no mental image."

There are two related reasons why it is hard to tell the story of Heisenberg's breakthrough into a new kind of quantum physics. One is that it involves a kind of algebra which is generally familiar only to pure mathematicians and quantum physicists and which even the quantum physicists have ways of avoiding; the other is that, like the story of Planck's quantum, it has been told in several different ways. Writers aiming at a report that is both accurate and comprehensible to the non-specialist reader find different ways of elucidating Heisenberg's thought processes. In so doing, they often disagree about minor historical details. This may not be their fault since in later years the characters

The Atom After Steiner — 183

involved seem to have had slightly different memories of the sequence of events. In order to give the reader an idea of how quantum physics was transformed in the years 1923-1927, I have drawn on many sources, including Max Born's Nobel Prize acceptance speech of 1954, my own experience as a student of atomic physics at Cambridge University in the 1950's, Gino Segrè's report in *Faust in Copenhagen*, in which he quotes Heisenberg at significant length and Helge Kragh's *Quantum Generations*.

The conceptual model of the atom, with its central nucleus supplied by Ernest Rutherford and its orbiting electrons contributed by Niels Bohr, had seemed very promising at first. As a general explanation of the properties of the elements and the meaning of the Periodic Table it was moderately successful, but it was clearly only a first approximation and there had been new developments in the study of spectra that were "perfectly unintelligible" (to use another of Born's phrases) in terms of the Bohr atom.

As described in Chapter V, Section (ii), the birth defect of the Bohr atom was that in order to make it work, its creator had to suspend one of the fundamental laws of electromagnetism while keeping all the other classical elements of mechanics and electrodynamics that he needed. Bohr used a ramification of Planck's quantum theory to calculate orbits for his electrons and to allow them to emit electromagnetic waves only when falling abruptly into lower orbits; but in other respects the electrons would be subject to classical mechanical principles, including Newton's Laws of Motion. This mixture of principles made it very hard to calculate precisely what was going on in the atom, and the only thing that made progress possible was the Correspondence Principle, which Bohr began to use at an early stage of the development of his model.

In order to understand this better I want you to imagine that you have three cups, a jug of water, a container of fine table salt and a box of sugar cubes. Into one cup you pour some water and note how it fills the cup and forms a nice horizontal surface. Into the next cup you pour some table salt and observe that it tries to behave like a liquid but can't quite manage it. It seems to fill the cup pretty well but there's always a little mound just where the stream of salt is landing. Turning to the next cup you try to pour the sugar cubes but you find that they won't pour very well and form a jumbled mass in the cup with a lot of empty space between the cubes. The moral is that a solid can to some extent mimic the properties of a liquid if it is finely divided enough, but a coarsely divided solid has no chance. The water represents classical physics in which all the variables are continuous, like the pressure and the volume in Boyle's Law or the energy of an accelerating particle. The sugar cubes represents the kind of physics in which the variables can change only abruptly, like the energy of an electron in Bohr's atom. When you serve the tea no one gets 2.74 lumps – they get one, two or three, but nothing in between. According to Bohr, however, there is in the physics of the atom a situation corresponding to table salt with its tiny crystals, in which the quantum jumps are so small that the changes are *almost* continuous. This happens when the quantum numbers are very large and explains why the old-fashioned, classical laws work very well under most ordinary circumstances, where we are dealing with matter (if it exists!) in bulk and not with individual particles. Bohr therefore concluded that one way of checking the validity of any particular bit of quantum theory was to see what happened if you figured out how it would operate when the quantum numbers were very large and the quanta correspondingly small. Under those circumstances if it didn't approximate to the classical law, it was probably wrong.

This began in 1913 and, as Max Born put it in his Nobel Prize acceptance speech, "Theoretical physics lived on this idea for the next ten years... It was a question of guessing the unknown from a knowledge of the limiting case..." Einstein, as Born recalled, made a big step in applying probability theory to the events inside the atom, but basic problems remained and "The art of guessing correct formulae that differ from the classical formulae but pass over into them as demanded by the correspondence principle, was brought to considerable perfection."

It wasn't enough and, as Born says, "This period was brought to a sudden end by Heisenberg, who was my assistant at that time. He cut the Gordian knot by a philosophical principle and replaced guesswork by a mathematical rule. The principle asserts that concepts and pictures that do not correspond to physically observable facts should not be used in theoretical description."

This "philosophical principle" had been growing in several people's minds. In 1924, the year of de Broglie's great *éclaircissement*, Heisenberg's friend Wolfgang Pauli had warned Niels Bohr that the mental image of orbiting electrons could no longer be tolerated; the realities were the measurable quantities, the frequencies of spectral lines from which the differences in energy levels could be calculated, whereas Bohr's atom was the result of making a picture of something that could not be observed. Heisenberg agreed, and in the following year he laid the foundation for a mathematical theory of the atom that would involve only measurable quantities and dispense with the mental image of the orbiting electron. This was a major and very difficult step; the concepts of energy and momentum had originally been developed in connection with the motions of material particles, but now we had to dismiss the old-fashioned images while retaining the numbers that had seemed to be attached to them. It was a bit like the situation of the Cheshire Cat, who gradually vanished but left his grin behind; "I've often seen a cat without a grin", thought Alice, "but a grin without a cat! It's the most curious thing I ever saw in all my life!" There is, indeed, something of wonderland and the looking-glass about the new world that Heisenberg was entering.

(iv)
Discontinuities and Probabilities

The treatment of variables like the energy and radius of an orbit becomes very different when these quantities can change only abruptly or, as the mathematicians say, discontinuously rather than smoothly. The following very simple illustration, which is not intended to be an exact analogy, may help.

The mathematics of Boyle's Law is simple. By pouring mercury into the long side of a J-tube, we can trap some air in the short side, measure its volume, v, and pressure, p, and calculate the product, pv. These are continuous variables and within the dimensions of the apparatus we can give p any value that we like and find the corresponding value of v. We find that, within the limits of experimental error, when we multiply p by v we always get the same answer; p times v is constant and we write, $pv = k$.

Suppose, however, that we lived in a world in which the pressure can take only whole number values and is adjusted by adding discrete weights to a platform resting on a piston. Our equation would be invalid except when p was a whole number and it would be necessary to find a different kind of mathematical formulation. This might be a chart of possible values or a graph showing a sequence of dots instead of a continuous line. Now imagine that the apparatus is enclosed in a large box, that we change the pressure by pushing a button and read the volume on a dial attached to the outside of the box. Just to make things more difficult, the dial registers only *changes* in volume and we have to try to imagine the volume of the gas changing from one value to another without having passed through any intermediate value. The entities that we are dealing with have become invisible and we can get our data only from observations that bear no resemblance to pressure and volume. You might, in fact, be tempted to say, "Look, we have an idea that things like pressure and volume exist inside this apparatus but we can't actually observe them. Furthermore, since the traditional forms of algebra and calculus don't fit variables of this kind, let's just find a mathematical form that fits the numbers and forget the pictures."

This gives a very rough idea of the situation that faced theoretical physicists who were dealing with transitions between the possible stationary states of an atom. If we think of this in terms of an actual electron – a point particle in a geometrical orbit – we have the difficult task of trying to imagine the electron disappearing from one orbit and reappearing in another without having been anywhere in between. The relations between energy levels calculated from the lines in the hydrogen spectrum could best be depicted by rectangular charts rather than the equations of traditional algebra and calculus; or, in Born's slightly more formal language, "Heisenberg banished the picture of electron orbits with definite radii and periods of rotation because these quantities are not measurable. He demanded that the theory should be built up of quadratic [rectangular] arrays..." He then went a little further and added the notion that what we should be interested in is the *probability* of these transitions taking place.

This sounds like an excellent plan as long as you know how to handle rectangular arrays and you know what these probabilities are.

*

Although Bohr's electron-orbit model had made it possible to calculate the frequency of the light emitted when the electron falls into a lower orbit, it gave no idea of the likelihood that this event would actually happen. When we look at the spectrum of hydrogen we are collecting the light from a huge number of atoms, and if we want to calculate the intensity of a particular spectral line – how much energy is pouring into it – we need to know the probability of the transition in the atom that gives rise to it. People who study probability in high school sometimes come away with the feeling that it is a very difficult subject and that they have only a hazy idea of what probability is.

As a very simple reminder, suppose that you have sixty dice, each of which is a perfectly constructed cube, and you roll them all. You assume that the numbers 1 to 6 have equal probabilities of showing up, but you know that this doesn't mean that you will get ten 1's, ten 2's, ten 3's and so on. Probability doesn't work like that; the probabilities are exact but they are, after all, only probabilities and the results show statistical variation. If you did the experiment with 6,000 dice, however, you might justly expect that each number will turn up close to $1/6$ of the time. If this doesn't happen, you are more likely to suspect that there is something wrong with the dice than that there is something wrong with the laws of probability. This may give us the uneasy feeling that the laws of probability are not subject to experimental verification – but that's another story.

Now to take a different kind of example; suppose you roll three dice – what is the probability of obtaining treble six? This is found by comparing the number of ways in which treble six can show up with the total number of possible outcomes. There is only one way in which treble six can appear – all three dice have to show a six. But the total number of possible outcomes is 6 times 6 times 6, which comes to 216. So the probability of a treble six is 1/216 and the same applies to any treble. Now, how about the probability of scoring 17? This can only be done with two sixes and a five, but it doesn't matter in which order the numbers appear, so there are three possibilities: 6, 6, 5; 6, 5, 6 and 5, 6, 6. The probability of scoring 17 is therefore 3/216 or 1/72. A little thought will show you that there are six ways of scoring 16 and that 10 and 11 are the most probable scores. We must remember, once again, that these precisely calculated probabilities can be seen to operate in practice only when we are working with large numbers of events, and that the results are always subject to statistical variation.

Probabilities can be calculated from the mechanical nature of the set-up and from the numbers of possible equally probable outcomes; but electrons, of course, are not dice. Einstein's work on the more remote and complex problem of the probabilities of possible electron transitions gave some useful results but, as Born reported, still left something to be desired.

To see what this has to do with rectangular arrays, all you have to do is to look at the chart at the end of a road atlas, in which the names of all the towns are written twice; along the top and down the left side. Each entry in the chart shows the distance from the town at the head of the column to the town at the end of the row. The diagonal from the upper left to the lower right is left blank because it represents the distance from a town to itself – the distance traveled if you stay where you are. Such an array is known in mathematics as a matrix, and, instead of towns, the kind of matrix that Heisenberg was looking for would have the energy levels of the electrons written across the top and down the side. The entries, instead of showing the distance between two places, would be the probabilities of the electron making the trip from one level to another. Students of High School algebra learn the rules for handling regular variables like Boyle's p and v, but Heisenberg needed the rules for operating with matrices and matrix algebra was a rather obscure branch of mathematics with which he lacked familiarity; so making some sense out of such arrays was quite a struggle.

(v)
HBJ or the Three-Man Paper

In May of 1925, two months after Rudolf Steiner's death, the twenty-four-year-old Heisenberg spent several weeks on the island of Heligoland in the North Sea, recovering from a severe attack of hay fever. The sparsely vegetated island with red sandstone cliffs, less than half the size of New York's Central Park, lies about thirty miles from the German coast and has the great advantages of a mild climate and almost total freedom from pollen. As Gino Segrè describes it, Heisenberg took only "a few items of clothing, a pair of hiking boots to climb the seaside rocks, a copy of Goethe's *Poems of the West and the East* and some calculations he was having difficulties with."

As the young physicist wrote later:

> *There was a moment in Heligoland in which the inspiration came to me.... It was rather late at night. I laboriously did the calculations and they checked. I then went out to lie on a rock looking out at the sea, saw the sun rise and was happy.*

Heisenberg had hit on the mathematical method and obtained some encouraging results, but he didn't immediately realize the full implications of the step he had taken, so he sent his work to Max Born and went to a previously scheduled conference in Cambridge. Born sensed that Heisenberg had put his finger on something fundamental, but it took him "a week of intensive thought and trial" to realize that Heisenberg had been wrestling with a form of matrix mathematics that he, Born, had known since his student days but hadn't thought about for many years.

The mathematics may have been old, but the physics was new and original. The mental image, as Heisenberg and Pauli insisted, had been discarded, the concept being expressed in purely mathematical form; and to anyone brought up on the traditional disciplines of algebra and calculus, the mathematics must have seemed very weird. In ordinary algebra the product of two or more numbers is the same whatever the order in which we multiply them; 3 times 2 equals 2 times 3. Multiplication is said to be *commutative*. Putting it algebraically we have

$$ab = ba$$

The rules for matrices are different. To the non-mathematician the idea of multiplying two matrices may seem incomprehensible – what do you get when you multiply two mileage charts? There are, however, definite rules for matrix multiplication and one of its oddities is that it is not generally commutative; in other words, for a matrix A and a matrix B, A times B is not the same thing as B times A.

Born applied the matrix multiplication rule to Heisenberg's quantum condition, found that it worked and also found a new reason for avoiding the mental image of an orbiting electron. Using the usual notation, in which q represents position and p represents momentum, p times q is not equal to q times p.

$$pq \neq qp$$

Since the two products are not equal there ought to be a calculable difference between them. Born did his calculation and "immediately there stood before me the following strange formula:" (I quote this to emphasize its strangeness, not because it will convey much to the non-mathematician.)

$$pq - qp = h/2\pi i$$

By that time in his life Born was used to the strangeness of quantum physics, but even he thought the formula was weird. He pointed out that this meant that the position and momentum of the particle could not be represented by ordinary numbers but by symbols whose product depended on the order of multiplication. It shows not only the non-commutative property of position and momentum but also the involvement with Planck's Constant, h, and the imaginary square root of -1, i. To a physicist, momentum (p) means mass times velocity, so this inequality implies that there is some difficulty in the idea of simultaneously knowing the position and velocity of a particle. Another two years of turmoil would elapse before this difficulty was given precise formulation in the shape of Heisenberg's famous Uncertainty Principle.

Born's excitement "was like that of a mariner who, after long voyaging, sees the desired land from afar, and my only regret was that Heisenberg was not with me. I was convinced from the first that we had stumbled on the truth. Yet, again, a large part was only guesswork..."

Born worked with his pupil, Pascual Jordan, to confirm the validity of Heisenberg's guesswork and "There followed a hectic period of collaboration among the three of us, rendered difficult by Heisenberg's absence... The result was a three-man paper which brought the formal side of the investigation to a certain degree of completeness."

The Three-Man-Paper was published in the summer of 1925, a few months after Rudolf Steiner's death, and there was, for the first time, a fully worked out quantum theory of the hydrogen atom. It was not greeted with universal enthusiasm for, like most people, physicists love their mental images and are apt to resent having to learn new forms of math. They would have been overjoyed if someone had come up with a different way of doing things, a way that enabled them to keep their beloved differential equations, work with continuous variables and eliminate the spectre of probability. Strangley enough, someone did and they were – at least for a while.

(vi)
Schrödinger's Wave Mechanics

The HBJ method of dealing with quantum physics wasn't the only show in town. When the three-man paper was published in 1925, Erwin Schrödinger (1887-1961) was already working on an alternative way of thinking about particles and waves – one which he hoped would enable us to keep a mental image of the inner workings of the atom, avoid such "monstrous" undesirables as quantum leaps and probabilistic interpretations and retain some version of the old concepts of continuity, waves and particles. His method, which took de Broglie's wave-particle duality as a starting point, gave rise to a discipline known as *wave mechanics*. A particle was still a particle – in this case, an electron – but associated with it was a wave that acted as a pilot or guide, and inside the atom the electron's wave had to fit exactly into an orbit. This concept resulted in energy levels that were exactly the same as those calculated by Bohr and had the advantage of employing traditional mathematical methods, supplying a visual image for people who needed one and keeping some form of the notion of cause and effect. Schrödinger liked his method and so did many other physicists, so it was a bit of an anticlimax when it turned out that wave mechanics was fundamentally identical with HBJ's matrix mechanics and did not, in fact, eliminate the quantum mechanical bogeys that its author disliked so much. It did, however, eliminate to a large extent the necessity for physicists to study matrix algebra. It's worth noting that in Richtmyer and Kennard's highly regarded *Introduction to Modern Physics* (McGraw-Hill, 1947), a chapter of 53 pages was devoted to wave mechanics while matrix mechanics was barely mentioned.

The strength of the friendship between Bohr and Heisenberg was severely tested by differences of opinion about Schrödinger's wave mechanics. In 1926, when it looked as if the tide of opinion was turning in his favor, Schrödinger visited Copenhagen, and he and Bohr spent a whole day battling over the possible representations of atomic structure. Eventually Schrödinger became sick and had to go to bed, but this didn't prevent the debate from continuing.

Bohr came to the view that there must be a synthesis that would include Schrödinger's pilot wave, but Heisenberg couldn't tolerate the pilot wave at any price, regarding it as just another mental image of something unobservable. Bohr and Heisenberg had spent many hours in exhausting but rewarding discussions of quantum theory, but now their talks became so heated and emotional that eventually Bohr went off on his own and a cooling-off period ensued. As

Heisenberg described the situation:

> Both of us became utterly exhausted and rather tense. So Bohr decided in February 1927 to go skiing in Norway and I was quite glad to be left behind in Copenhagen, where I could think undisturbed about those hopelessly complicated problems.

Meanwhile, in a paper published in 1926, Max Born had shown that Schrödinger's pilot wave could be interpreted as a probability wave, the square of its amplitude being a measure of the probability of finding the particle within a given cell of space-time. According to this interpretation, there is no precise description of the present, and the only precise information leading into the future is a set of probabilities. A precise knowledge of probabilities is a very different thing from a precise knowledge of future events. Born belatedly received the Nobel Prize for physics in 1955, but his paper, dealing as it did with the future in terms of probabilities rather than calculable motion, was pleasing neither to Einstein nor to Schrödinger. Realizing that in spite of all his efforts he had not been able to restore causality and continuity to atomic physics, Schrödinger expressed acute disappointment and said that he wished he'd never had anything to do with it. He was, however, a very resilient and unconventional character, who surprised his new Oxford colleagues in 1933 by showing up with wife *and* mistress in tow. A whole new institute was created for him in Dublin in 1939 and he was an honored and popular figure for the rest of his life.

After Bohr's departure, Heisenberg put his thoughts in order to such good effect that he was able to crystallize the principle of uncertainty or indeterminacy that some have described as the fundamental generalization of quantum physics.

(vii)
Indeterminacy

Electrostatics is a favorite subject in elementary science courses. Many of us have seen experiments with charged pith balls, Leyden jars, electroscopes and Wimshurst machines, so we are familiar with the idea of a material object carrying an electric charge. As J. J. Thomson had remarked, "The assumption of a state of matter more finely divided than the atom of an element [was] a somewhat startling one," but he and his colleagues soon got used to the idea of the electron and gave it the same mathematical treatment as any other charged object. The electron, however, had other ideas, and as the years went by it became more and more elusive, diddling the observers by behaving sometimes as a particle and sometimes as a wave. So Heisenberg and his colleagues concluded that enough was enough and decided to stop visualizing the electron as either a particle or a wave – in fact, to stop visualizing it.

History and personal experience indicate, however, that it's very difficult, if not impossible, to get rid of such mental images. I am told that Owen Barfield suggested that a good exercise would be to go into a corner and not think about pink elephants. Wishing to retain whatever sanity remains to me, I have not tried this, but as a physicist I have been forced to contemplate the possibility of thinking about electrons without thinking about them or, at least, talking about them without mentioning them. This is the problem that makes it next door to impossible to give an accurate explanation of Heisenberg's Uncertainty Principle[99] in anything except purely mathematical terms.

To put things somewhat loosely, a *particle* has a definite location within a chosen frame of reference – it's always *somewhere* – and it's always doing something or nothing.

Maybe it's moving in a certain direction at a certain speed or maybe it's at rest. Physicists like to be very specific about particles, stating their exact positions, speeds and directions of motion; but Heisenberg's way of thinking (or not thinking) about electrons made this impossible. According to the *Concise Dictionary of Physics*, the principle states that it is not possible to know with unlimited accuracy both the position and momentum of a particle. (If you are not familiar with the physicists' concept of momentum as "mass times velocity" you can get a good idea of what Heisenberg's principle means by simply substituting "velocity" for "momentum.") Applications of this principle are usually about electrons, but it applies to any particle.

The dictionary statement seems to suggest that the particle actually has a definite position and a definite momentum but that if we try to measure them both at the same time, we have to be content with an approximate result. Increasing the accuracy of one measurement necessarily decreases the accuracy of the other. In talking like this we have done exactly what Heisenberg said we shouldn't do. Bowing to the necessity of communicating something intelligible, we have talked about the particle as though it were an ordinary object with ordinary mechanical characteristics, like Bohr's orbiting electrons.

One way of deriving the principle keeps the particle image and involves the nature of seeing – at least, as usually understood. If you want to know where something is, you look at it. Sunlight or lamplight bounces off the object and enters your eye. The object may be a No. 7 bus approaching you down Columbus Avenue, a spinning baseball approaching your head at 97 miles an hour or your cat sleeping in front of the fire. The fact that you are looking at the bus or the baseball makes no discernible difference to its progress. Matters may be different with the cat, since cats always seem to know when someone is watching them, but that only shows that there are aspects of vision that fall outside the physicists' chosen domain. In physicists' language, which is perfectly appropriate for their methodology, the photons that bounce off the bus before entering our eyes have so little momentum that their effect on the vehicle is immeasurably small. What happens, however, if you try to look at an electron?

The photon, which was so puny in relation to the bus, has a degree of heftiness on the same scale as that of the electron, so whatever the electron was doing before the photon hit it, it's doing something quite different now. In principle, you may be able to get some information about an electron by bouncing a photon off it, but by the time you get it, it's past history and goodness only knows where the electron is now and what it's doing. The degree to which the electron can be pinned down by this method can be expressed mathematically, and it agrees exactly with the dictionary statement of Heisenberg's Principle. Max Born in his *Atomic Physics*, gives several varieties of such mental experiments for pinning down the electron, like letting it pass through a tiny aperture, or looking at it through a microscope. All of these mental experiments give the same result; as the *Concise Dictionary* states, there is an uncertainty of location and an uncertainty of momentum, and the product of these two uncertainties has a minimum value that is a multiple of Planck's constant. The more exactly we know the particle's momentum, the less exactly we know where it is, and *vice versa*. Born goes into these matters at some depth, but he saves the real McCoy for an appendix.

The real McCoy, which Born calls the "rigorous derivation," is usually reserved for post-graduate students, and it doesn't say anything about electrons or particles, being based on the fundamental nature of the problem rather than the difficulty of coming up with a suitable technique. The real problem is not that we haven't yet thought of an experimental setup for measuring the position and motion of an electron, but that *an electron doesn't have a precise position and motion.* You can't measure something that isn't there. Anything that has a precise position and motion is an old-fashioned particle, and an electron isn't an old-fashioned particle. It isn't a wave, either, and it isn't something in between. This makes it very hard to talk about, since whenever we say "electron" we're apt to imagine a tiny charged particle. When I was a student I gathered that Heisenberg's Principle was about the impossibility of simultaneously knowing the exact position and exact momentum of an electron, but actually the principle doesn't mention electrons, although we always seemed to be talking about them. Since the Uncertainty Principle arises from a purely mathematical, non-visual treatment of conditionally measurable quantities, it can be stated only in mathematical language. In Heisenberg's words – and I hope you're sitting down and ready for this – "the quantum uncertainty in the simultaneous determination of both members of a pair of conjugate variables is never zero."

If we were in a classroom, this would be the point at which someone would say, "Do we have to know this?" or "Is this going to be on the test?" and I would say, "No, my dears, this is merely to show you how far atomic physics had traveled in the decade or so leading up to 1927." In 1917 we thought its aim, partially achieved, was the exact description of the activities of particles and waves; but according to Heisenberg, Born, and Jordan, its content is really the transformations of mathematical entities that most physicists had never heard of at the time. And, as Heisenberg showed, it isn't merely that it is in principle impossible to know exactly where a particle is and what it's doing but that the whole idea is meaningless.

If this is so, you may well ask why it is that physicists talk about atoms, electrons, protons, photons, pions, muons, gluons and quarks, all the time, not to mention the fearsome Higgs boson (the Loch Ness Monster of modern physics.)

One obvious reason is that a lot of the time it works quite well to think in terms of particles and waves and another is that it makes communication possible, especially to the non-specialist. It's true that there are sub-atomic phenomena that remain obstinately paradoxical when thought of in pictorial terms, but for most practical purposes, especially among electrical engineers, chemists and biologists, electrons and waves do the job very well. By "practical purposes" I mean such things as explaining how transistors work, why sodium and potassium have similar chemical properties and how energy is made available in muscle tissue. It's really only when you get right down to bedrock that the old images create such problems that they lose their power to generate and communicate insights and, in Max Born's words, we realize *"the hopelessness of trying to account for the properties of elementary particles in terms of simple mechanical models."*

*

The point that was of the greatest interest to the philosophers and theologians of the time was that Heisenberg's Principle seemed to rule out the kind of determinism put forward by Laplace[100] in the early nineteenth century, according to which the exact state[101] of the world at any moment determines its exact state at any future moment. This idea could be seriously upheld by anyone who believed that the world consists only of atoms and that the motions of the atoms are completely determined by precise mechanical laws. If, however, the world is made of entities that are neither waves nor particles and have no precise state of location and motion, we can never specify the exact state of the world and this form of determinism seems to be untenable. I have to say, "seems to be" because determinism was a concept which some philosophers were very loath to give up, and the Laplacian form got tangled up with philosophical-religious forms, resulting in the expenditure of great quantities of ink and air.

(viii)
Quantum Physics and the Periodic Table

When the *Knaben* did away with the mental image of the orbiting electron they didn't totally destroy the notions of shells and energy levels that had made such a good start on the explanation of the Periodic Table [See *Ch. V(iii)*]. What happened was that the idea of an electron proceeding around a circular or elliptical path like a planet in its orbit, was replaced by the concept of a symmetrical region of space around the nucleus, in which there was a finite probability of the electron making its presence felt at any given point and moment. The shapes and sizes of these regions and their probability distributions can be calculated and, as you would expect, the regions (*orbitals*) for electrons at different energy levels overlap, just as the old-fashioned shells had overlapped. The old method was in many circumstances an excellent approximation for the new one, but the latter made it possible to get the right answers to problems, such as those of the helium atom and certain effects of electric and magnetic fields, that had caused so much alarm and despondency before.

While the young Turks and their elders were busy refining the concepts of the new quantum physics, an equally young American chemist was exploring their application to the problems of chemical bonding. Starting in 1931, Linus Pauling (1901-1994), regarded by many as the greatest chemist of the twentieth century, produced a series of books in which he more or less invented what became known as "quantum chemistry," explaining how the orbitals determined the chemical properties of the elements and accounting quantitatively for the different kinds of chemical bond. One of Pauling's achievements was to show that the carbon atom has tetrahedral symmetry. Its four bonds are distributed spatially like the lines joining the center of a regular tetrahedron to its vertices. This, together with the actual size of the atom, is what makes the extraordinary proliferation of organic compounds possible. By 1940 it was generally felt that the mostly orderly and occasionally capricious relationships of the elements in the Periodic Table were very well understood.

Also in 1931, Paul Dirac produced a new equation for the electron which included Heisenberg's and Schrödinger's methods as special cases and demonstrated their equivalence. At this point the transition from the old quantum physics to the new was substantially complete.

(ix)
More about Probability

If atomic physics at the deepest level is not about particles or waves, what is it about? Or, to put a slightly different question, when does an atom make a physicist sit up and take notice? The answer to the second question is, "when it emits or absorbs a flash of light." The physicist calls the flash of light a photon. When we try to imagine what's happening

The Atom After Steiner — 203

inside the atom, even an old-fashioned Bohr atom, our imaginations fall short because we can't picture an electron getting from one orbit to another without traversing the intervening space. We can, however, detect the flash of light. It's an event, and now we know the answer to the first question.

Particle physics as developed by HBJ is not about particles but about *events*, although to hear the way physicists talk you would never think so. The real difference between an electron and a baseball is that the properties of an electron are expressible only as a set of mathematical rules governing the things that might happen to it – events – whereas a baseball has form, texture, elasticity and internal structure as well as momentum and energy. This is not a criticism of the electron, a concept which serves its purpose admirably. If it had anything beyond mathematical rules for its transitions attached to it, it would not be a fundamental entity. In Bohr's words, *"Isolated material particles are abstractions, their properties being definable and observable only through their interaction with other systems."*[102] The only time we can be aware of an electron is when it does something or has something done to it, and by the time we have became aware of it, it's utterly elsewhere doing something else. We can see the track of an electron in a cloud chamber, but what we see is really the trail left by successive events, not the electron. Meanwhile, our baseball, when not in play, can be observed sitting serenely in a box with a bunch of other baseballs.

As Max Born had shown, quantum and wave mechanics made it possible to calculate the *probabilities* of events on the subatomic scale. Unlike the position and momentum of a particle, the probability of an event could be calculated exactly. Then, however, we are left with the fact that probabilities tell us nothing about individual objects and what happens to them in particular cases, but only about the average outcomes of large numbers of events. There is no way of telling at what moment a particular electron in a particular atom will make its move, but the overall behavior of huge numbers of electrons in huge numbers of atoms can be predicted with great accuracy. Maxwell and Boltzmann had already applied statistical methods to the atoms and molecules of the old kinetic theory, and it had been observed that the decay of radioactive elements appeared to obey statistical rules; Planck had used statistical theory to develop his quantum theory and now the statistical interpretation looked like the only way of dealing with subatomic physics.

The idea that the predictable working of cause and effect in the physical world might be based ultimately on a sub-world of discontinuity, probability and indeterminacy didn't go over very well with some physicists, including Einstein, whose comment that God doesn't play dice has been quoted many times in several different forms. "I don't want to let myself be driven to a renunciation of strict causality before there has been a much stronger resistance to it... I can't bear the thought that an electron exposed to a ray should by its own free decision choose the moment and the direction in which it wants to jump away," he wrote to Born in April 1924. "If so I'd rather be a cobbler or an employee in a gambling house than a physicist." Two years later, referring to the content of the Three-Man Paper, he wrote, "Quantum mechanics is most awe-inspiring, but an inner voice tells me that it is not the real thing after all. The theory gives much but it scarcely brings us nearer to the secret of the Old Man. In any case, I'm convinced that He doesn't play dice."[103] The correspondence on the subject between Einstein, Born and Bohr is a study in itself, but here it will have to suffice to say that although the probabilistic interpretation has been repeatedly challenged and expressed in different mathematical forms, it remained for many years the most widely accepted one.

*

People with a decent background in physics could quite easily read Heisenberg's chaste statement of the Uncertainty Principle without having the faintest idea of what it means or what it has to do with atomic structure. I quoted it because the situation is not without an element of humor and because it shows how remote the principles of quantum physics had become from human sensory experience. This is evidently not the kind of mathematics that Steiner had in mind he when spoke of bringing "order and harmony into the otherwise chaotic stream of everyday facts..." Unlike the arithmetic of Boyle's Law, which deals with the visible or tangible properties of volume and pressure, quantum physics has no apparent relationship to everyday facts. We experience the reality of the gas laws every time we pump up a tire, but there is nothing in our everyday experience that prepares us for the wave-particle duality and the uncertainty principle. Quantum physics is basically uninterested in everyday experience, drawing its data from experimental set-ups that bear no resemblance to anything that appears in the daily round and the common task. Reading casually about the Compton effect, you might come away with the impression that someone had actually seen an X-ray photon collide with an electron. We have to remember that we have only "seen" it inferentially.

One may well "admit that all the knowledge obtained in this way stands as a closed door to the outer world in that it does not allow the essence of this outer world to enter our cognition..." but it must be recognized that the physicists reached this point in the course of their quest to *find* the essence of the outer world and that it is a very different kind of essence. In Goethean science the essence is the Archetype revealed through contemplation. In modern physics the essence is a teeming world of invisible particles that become less and less substantial the more profoundly we investigate them.

"As long as we remain in this field of knowledge [quantitative science], we cannot see through the outer appearances; we also, of course, do not claim to do so..." But to see through the outer appearances is exactly what atomic physicists and quantum physicists attempt to do. Like Parmenides, they might well argue that the natural, everyday world of appearances is an illusion, but, unlike the great pre-Socratic, they would assert that beneath the appearances is a world of incessant change, lacking even the permanent, substantial atoms of Democritus, Newton, and Dalton. We have already seen, however, that if we enthusiastically feel that in abandoning old-fashioned atomic physics, Heisenberg and company had successfully dismissed the atomic particles generated by the mental inertia, as Steiner had seen it, of previous generations, we are in for a disappointment. Particles are still with us in ever increasing numbers but there does seem to be a Correspondence Principle for them. Little ones, like the electron, are governed by quantum rules that make it hard for us to think of them as particles; big ones, like the proton, are governed by the same rules but their size usually makes it possible for us to treat their behavior according to everyday mechanical and electrical principles.

(x)
Niels Bohr – a Goethean Physicist?

The poet Wordsworth wrote of "Newton with his silent face," his "mind forever voyaging through strange seas of thought alone." This highly idealized picture of the idiosyncratic genius has the effect, not, perhaps, intended by the poet, of creating an impression of the scientist as cool, detached, dispassionate, and slightly inhuman. The people we're talking about were independent, cooperative, flesh-and-blood individuals who were devoted to science and deeply interested in many other things. The one who least fits Wordsworth's picture of the scientist is the voluble, gregarious, sport-loving Niels Bohr.

Bohr, as you know, was in Manchester with Rutherford in 1911 shortly after the discovery of the nucleus. He returned to Copenhagen in 1912 but in 1914 he accepted a readership in theoretical physics in Manchester. This caused his fellow countrymen such anxiety that in 1916 they offered him a full professorship and an institute of his own if he would return to his home town. This was an offer he couldn't refuse and in 1918 he became the first director of Copenhagen's Institute for Theoretical Physics. This sounds very grand, but the reality was that in the beginning Bohr shared a tiny office with his young Dutch assistant, Hendrik Kramers, and a secretary hired the following year. That the Institute was able to move into a more spacious building was due largely to Bohr's energetic fundraising activities, which he somehow managed to combine with his continuing adventures as a theoretical physicist.

By the end of the First World War, Bohr's work on atomic structure had made him so famous that younger physicists from Europe and, later, America, flocked to Copenhagen. What kept them coming back, year after year, was not only

Bohr's prestige and expertise but also his warmth of heart and wonderful hospitality. When Heisenberg wrote, "Bohr's influence on the physics and the physicists of our century was stronger than that of anyone else, even that of Albert Einstein," he was referring less to Bohr's intellectual achievements than to the manner in which his style "affected the way physicists think and work... how they relate to their mentors, their peers and their students." Because of this, Bohr "became the most loved theoretical physicist of his generation."[104]

"Yes, loved," Segrè emphasizes. Young physicists respect and admire their great contemporaries and predecessors, "but love is something different. Yet it is a term that appears again and again in memoirs when physicists speak of Bohr."

Unselfish giant that he was, Bohr devoted much time and energy to helping others through all kinds of difficulties. During the 1930's, when the Nazi Party was tightening its stranglehold on German society and eliminating "Jewish" science, and during the early years of the war, he provided a refuge for many scientists fleeing persecution. His mother was Jewish, so when Copenhagen was overrun by the German army in 1943 he recognized that it was time to leave. With his wife and a few friends he escaped to Sweden by night in a fishing boat. He immediately asked the king to let it be known that Sweden would provide a refuge for Danish Jews and within two months most of Denmark's Jews, including Bohr's family, were rescued by the same route. Perhaps most revealing of Bohr's personality is his long, difficult, and loving association with Werner Heisenberg.

Heisenberg first met Bohr in Göttingen in 1922 when he and Pauli attended the series of lectures in which the Dane expounded his theory of the Periodic Table. It may be only a slight exaggeration to say that the two young men, whose ages at the time added up to forty-two, were bowled over. As Heisenberg said, "We had all of us learned Bohr's theory from Sommerfeld and knew what it was about. But it all sounded quite different from Bohr's own lips."[105] Segrè expresses the matter in musical terms: "As with a musical virtuoso, whose performance is more than just the playing of notes, the phrasing and the emphases mattered. One learned from Bohr more than simply what the equations said."

According to Heisenberg:

> Bohr's insight into the structure of the theory was not the result of a mathematical analysis of the basic assumptions, but rather of an intense occupation with the actual phenomena, such that it was possible for him to sense the relationships intuitively, rather than derive them formally. Thus I understood: knowledge of nature was primarily obtained in this way, and only as the next step can one succeed in fixing one's knowledge in mathematical form and subjecting it to a complete rational analysis. Bohr was primarily a philosopher, but he understood that natural philosophy in our day and age carries weight only if its every detail can be subjected to the inexorable test of experiment.

Hendrik Kramers gives a similar message in more concrete terms:

> It is interesting to recollect how many physicists abroad thought at the time of the appearance of Bohr's theory of the periodic system, that it was extensively supported by unpublished calculations which dealt in detail with the structure of the individual atoms, whereas the truth was, in fact, that Bohr had created and elaborated with a divine glance a synthesis between results of a spectroscopical nature and of a chemical nature.

Kramers recalled the way many physicists perceived Bohr's

semi-intuitive use of the Correspondence Principle:

> In the beginning the Correspondence Principle appeared to the physicists as a somewhat mystical magic wand that worked only in Copenhagen.[107]

Comparing Bohr with the rigorously analytical Sommerfeld, Segrè points out that instead of precisely formulating the problem before seeking a solution, Bohr "struggled simultaneously toward defining a question and answering it" and tried to get his audience involved in the search. What Segrè does not suggest, although he may well have thought it, is that this is another facet of the objective-subjective polarity; in Sommerfeld's case, "A carefully defined question is in me and somewhere out there is an answer that fits the question." And for Bohr, "The question and the answer grow together through the interplay of inner and outer."

*

Many of Bohr's colleagues found his methods deeply disturbing but had to admit that they often worked. If, like Heisenberg, we speak of someone whose "intense occupation with actual phenomena" was "such that it was possible for him to sense the relationships intuitively," we can't help thinking of Goethe. Furthermore, when we add that the next step was to "succeed in fixing one's knowledge in mathematical form," we recall Steiner's questions:

> Could the force which we have to use to attain a mathematical knowledge of nature be used more effectively, with the result not just of a mathematical abstraction, but something inwardly, spiritually concrete? ...This we can see as a third step in attaining knowledge. The first step would be the familiar grasping of the real outer world. The second would be the mathematical penetration of the outer world,

after we have first learned inwardly to construct the purely mathematical aspect. The third would be the entirely inner experience, like the mathematical experience but with the character of spiritual reality.

So wouldn't Steiner have eagerly embraced this new kind of fundamental physics? Doesn't it look as if Bohr had accomplished the first two steps in a thoroughly Goethean manner that would have met with Steiner's entire approval? If we consider the kinds of phenomena Bohr had contemplated, we may conclude that the answer is, "Probably not." Goethe had generally allowed nature to speak to him without coercion, whereas the input to Rutherford's, Bohr's, and Heisenberg's theories came from nature "constrained and vexed." Rutherford bombarded thin metals foils with alpha particles from radioactive sources; the objects of Bohr's contemplation were things like the frequencies of the spectral lines, obtained by applying high voltages to highly rarefied gases, and the structure of the Periodic Table, also the result of a great deal of constraint and vexation; Heisenberg's mathematics came as the result of a struggle for a new and more consistent interpretation of the same kinds of data. And yet, Bohr's colleagues evidently felt that there was something of the mystic about him, almost as if he sometimes obtained his results from a region to which they had no access.

To anyone who is tempted to remark, with a knowing look, "Well, we know where that was," I would say, "Hold on – don't go so fast." As Steiner says, Ahriman has a foothold in everyone's soul ("We are all tainted with atheism") but my impression is that Bohr's soul was unusually healthy, and that his commitment to finding as much of the truth as he could understand was intimately connected with his warm and compassionate heart. He was, in fact, a good exemplar of the kind of selflessness that the times demand. He was responsible for a major step in what Steiner called the electrification of the atom, and he was extremely enthusiastic about his work; but he was able to see and acknowledge its deficiencies and play an indispensable part in the move to a new kind of physics, one that any anthroposophist who is interested in science has to take very seriously and try to understand.

(xi)
Are Atomic Particles Real?

How can we sum up the fate of the old notions of atoms and atomic particles? One of the results of this brief sketch of events in the world of atomic physics in the immediate aftermath of Steiner's death is a question, not an answer. Here it is, as posed by Max Born in his Nobel Prize speech of 1955:

> *So now we come to the final point – can we call something with which the concepts of position and motion cannot be associated in the usual way a thing, a particle? And if not, what is the reality that our theory has been invented to describe?*
>
> *The answer to this question is no longer physics but philosophy... Here I will only say that I am emphatically for the particle idea. Naturally it is necessary to redefine what is meant...*

It would be a fine stroke of rhetoric to leave this last sentence hanging and go on to the next chapter, but it would be unfair to Born, who was a great physicist, a fine teacher and an admirable human being. In fact, he explained exactly how the adjustment could be made using the mathematical concept of invariance with respect to transformation – determining what it is that remains constant while all else changes. The significant things are that the question had to be asked, that it took such a lot of work to answer it and that Born was already aware in 1955 that the answer had its limitations.

The latest research on nuclei and elementary particles has led us, however, to limits beyond which this system of concepts itself does not appear to suffice. The lesson to be learned... is that probable refinements of mathematical methods will not suffice to produce a satisfactory theory, but that somewhere in our doctrine there lurks a concept, unjustified by experience, which we must eliminate to open up the road.

In the fifty-seven years that have elapsed since Born made these remarks, we have acquired a lot of new particles, including a whole extended family of quarks, and some new varieties of fundamental theory, including string theory and quantum gravitation; but we don't seem to be much nearer to the physicists' El Dorado, the "theory of everything." This would not only "predict" the existence, however shadowy and unimaginable, of all these particles and the forces – electromagnetic, gravitational and nuclear – that play on them, but also indicate experimental methods by which it could be verified. Born had the idea that the way forward would be to search for and dispose of some further unobservable factor, a "poison pill" still lurking in our theoretical framework. More recently, however, the tendency has been to introduce more unobservables rather than to try to eliminate hitherto unsuspected ones. String theory, for instance, brings in so much that is unobservable that it is capable of generating millions of possible worlds without being able to make any significant verifiable predictions or even to specify which, if any, is (or are) the one (or ones) we live in.[108]

My impression is that behind all the efforts to explain existence in terms of hidden variables, strings, multiple worlds, or whatever imagined solution seems as if it might do the trick, is the ineradicable desire for a mental image. And these mental images often work up to a point – there is something quite satisfying about the picture of three quarks making a proton – but there are always ramifications, as if the concentration of thought into a crystalline image creates a small region of peace, quiet, and order at the expense of chaos in the rest of the universe. One way of dealing with this situation is then to declare that the rest of the universe, namely the world of consciousness, life and spirit, is an illusion. The obvious problem with this opinion – that the consciousness of the person stating it is therefore an illusion – is not often mentioned.

It is my deeply considered opinion that Bohr and the physicists working with him were justified in using quantum mechanics as a practical method of proceeding with their tasks even when they didn't always understand why it worked. This procedure became known as "quantum cookery," and the general tendency today is to disapprove of it. There is, however, something very down-to-earth about it – an attitude of working with the materials that we actually have before us, that appeals to the Goethean side of my make-up, in spite of the ill-treatment of nature that lies behind it.

Chapter VIII
Epilogue

As far as the scientifically-minded anthroposophist is concerned, one of the most appealing features of the quantum physics of the 1920's and '30's may well be the effort to manage without mental images of the subatomic world. If, as some appear to believe, we have taken the intimate investigation of matter to the point where it touches on the world of living forces that Steiner identified as the etheric, we shouldn't be surprised to find that concepts appropriate to the investigation of what is dead in the material world no longer apply in the old ways. It is possible to believe that the paradoxical nature of some of the findings of quantum physics reflects the interaction of the physical and the etheric – the interplay between a world that we feel we know and one that is pervasive but invisible and whose operation is essentially mysterious. We may consider that our high-energy search for nature's inner secrets has taken us beyond the boundary of the physical and that our old images of material particles make it harder to perceive influences from a non-material source. Perhaps if we abstain from material images, while acknowledging the polarities of wave-like behavior and particle-like behavior, we may grasp to some extent the rhythmical working of the etheric.

I would not presume to tell anyone what to think, but there is no harm in uttering a mild caution, even when it involves repeating something that I have already said more than once. From the alchemists to Goethe and Steiner, people have had contemplative ways of perceiving the functioning of the etheric and, to some extent, of the worlds beyond it. My respect and admiration for many of the quantum physicists and their immediate forerunners does not prevent me from reminding you that their experimental methods could hardly have been less in tune with the ideals of Goethean science. Goethe objected to Newton's comparatively innocuous experimental technique of allowing sunlight to pass through a narrow opening, so what he would have thought of the vacuum pumps, induction coils and diffraction gratings of the later nineteenth century and the cyclotrons, bevatrons and super-colliders of the twentieth is not hard to imagine. Furthermore, from the earliest studies of the spectra produced in vacuum tubes to the high energy research of the late twentieth and early twenty-first centuries, everything has been electrically based. Bearing in mind all that Steiner said about the atom and electricity, we may well feel that, far from penetrating the etheric, modern quantum physicists have created a new, artificial world and that Ahriman has had an influential hand in its development. This new world, if such it be, is, nevertheless, the result of the persistent search for truth of many devoted workers. It interacts with the old world of matter and sense experience and, as I sit here at my computer, I must remember what Steiner said about "thundering against Ahriman." It doesn't do any good.

The question of what Goethe and Steiner would think if they were alive today is certainly hypothetical, but dwelling on it may help us in the crucial attempt to clarify what we think. If our thoughts are selfish, careless or casual, we easily become victims of Ahriman or his seductive colleague. If we develop our capacities to the point of being able to see and understand what's going on around us, we may have something vital to offer. In trying to do this we can find great help and inspiration not only in the life and work of Goethe and Steiner but also in the magnificent efforts of people like Planck, Einstein, Bohr, Heisenberg and Born to project their thinking into the invisible recesses of the physical world.

*

In a lecture on Karmic Relationships, given in Prague on April 5th, 1924, Steiner tackled one of the problems of anthroposophy – the question of its strangeness:

> *Even to people who, comparatively speaking, are kindly disposed, Spiritual Science still seems strange and foreign. One cannot read without a certain irony what someone, who is in other respects so promising, says about me as the founder of Anthroposophy. In* The Great Secret, *Maurice Maeterlinck*[109] *seems unable to deny that the introductions to my books contain much that is reasonable. He is struck by this. But then he finds things which leave him in a state of bewilderment and of which he can make absolutely nothing. — We might vary slightly one of Lichtenberg's*[110] *remarks, by saying: 'When books and an individual come into collision and there is a hollow sound, this need not be the fault of the books!' But just think of it — Maurice Maeterlinck is certainly a high light in our modern culture and yet he writes the following — I quote almost word for word: 'In the introductions to his books, in the first chapters, Steiner invariably shows himself possessed of a thoughtful, logical and cultured mind, and then, in the later chapters he seems to have gone crazy.' What are we to deduce from this?*

> First chapter — thoughtful, logical, cultured; last chapter — crazy. Then another book comes out. 'Again, to begin with, thoughtful, logical, cultured; and finally — crazy!' And so it goes on. As I have written quite a number of books I must be pretty expert at this sort of thing!
>
> According to Maurice Maeterlinck a kind of juggling must go on in my books. But the idea that this happens voluntarily... such a case has yet to be found in the lunatic asylums!
>
> The books of writers who think one crazy are really more bewildering still. The very irony with which one is bound to accept many things to-day shows how difficult it still is for men of the present age to understand genuine spiritual investigation. Nevertheless such investigation will have to come. And in order that we shall not have been found wanting in the strength to bring about this deepening of the spiritual life, the Christmas Foundation Meeting was held as a beacon for the further development of the Anthroposophical Society in the direction I have indicated. The Christmas Foundation Meeting was intended, first and foremost, to inaugurate in the Anthroposophical Movement an epoch when concrete facts of the spiritual life are fearlessly set forth — as has been the case to-day and in the preceding lectures. For if the spirit needed by mankind is to find entrance, a stronger impetus is required than that which has prevailed hitherto.

*

It is helpful to remember these words when we meet apparent contradictions and ideas that it would be much easier to dismiss as fantasies. Let me repeat some basic observations:

> 1. In public communications, Rudolf Steiner was a lifelong opponent of atomic theories. He regarded atoms as mental constructs.

2. In lectures addressed to members of the Theosophical Society and the Anthroposophical Society he referred to atoms as though they were physically real and spoke of the diabolical qualities that emerged when they were given electrical characteristics.
3. The atom that Steiner continued to reject publicly as late as 1923 was very different from the one that physicists put together in the period from 1925 to 1930.
4. The atom was originally a speculative response to a philosophical problem but latterly its development was driven by experimental observations made in a manner antithetical to the ethos of Goethean science.

We can safely assume that many of those present at the Members' lecture of January 28th, 1923 had also been present at the public scientific courses. One wonders how, in Steiner's actual presence, they experienced what, in cold print, seems to be a contradiction. Possibly some had been at the lecture of December, 1904, in which Steiner spoke of the reality of the electrical atom and linked electricity with thinking. Some, also, must have read the Goethe prefaces and been aware of Steiner's lifelong resistance to the atomistic interpretation of nature. To repeat a paragraph from the 1904 lecture:[111]

We are going forward to an age when men will understand what the atom is, in reality. It will be realized — by the public mind too — that the atom is nothing but coagulated electricity. — Thought itself is composed of the same substance. Before the end of the fifth epoch of culture, science will have reached the stage where man will be able to penetrate into the atom itself. When the similarity of substance between the thought and the atom is once comprehended, the way to get hold of the

> forces contained in the atom will soon be discovered and then nothing will be inaccessible to certain methods of working. — A man standing here, let us say, will be able by pressing a button concealed in his pocket, to explode some object at a great distance — say in Hamburg! Just as by setting up a wave-movement here and causing it to take a particular form at some other place, wireless telegraphy is possible, so what I have just indicated will be within man's power when the occult truth that thought and atom consist of the same substance is put into practical application.

And now, a paragraph from *The Boundaries of Natural Science*:

> It is interesting to note that a great proportion of the philosophy that does not remain within phenomena is actually nothing other than just such an inert rolling-on beyond what really exists within the world. One simply cannot come to a halt. One wants to think ever farther and farther beyond and construct atoms and molecules—under certain circumstances other things as well that philosophers have assembled there. No wonder, then, that this web one has woven in a world created by the inertia of thinking must eventually unravel itself again.

In 1904 Steiner spoke of atoms as if they were constituents of the physical world; this appears to contradict his perception that the atom is a mental construct of something that doesn't exist. It is possible to regard this contradiction both as a difficulty to be overcome and as a pointer to a deeper understanding. Alternatively, especially when we read of the electrical atom as a demon of evil, we might throw up our hands and conclude either that the whole thing is too much for us or simply assume, as Maeterlinck did, that Steiner was a nut-case. How we decide to proceed depends on previous experience, mine being that although Steiner is not infallible, his view of the big picture of world history and the intimate nature of the human being has repeatedly turned out to make sense and yield unparalleled insight and understanding. Let me, however, play the cynic for a moment.

I've already mentioned that in using the phrase "coagulated electricity," Steiner was giving a rough description of the contemporary physicists' picture of the atom as a bunch of electrical particles held together in some unknown fashion. It is also arguable that when he described thought as being composed of electricity, he was merely reflecting the prevailing view of the nervous system. At that time, there was great excitement over the marvels of long-distance electrical communication, so it wouldn't have been much of a step for Steiner to put these ideas together and produce a fragment of early science fiction, acceptable to committed followers but not suitable for an audience containing professional scientists.

These are very deep waters, and the object of the cynic is to make them as shallow as possible. Cynicism is very easy and very tempting; it requires hardly any effort, intelligence, or insight and confers an aura of smartness and sophistication on the exponent. The cynic is not always wrong, but justifiable cynicism is usually known as *realism* – not, of course, in the philosophical sense of the word. So what is to say that the view I have just given isn't the *realistic* one? It would, after all, be quite reasonable to suggest that a speaker or writer who frequently presents us with ideas that are shocking, contradictory, or excessively hard to understand might comfortably be ignored. *In that case, however, we might have to ignore the quantum theory and wave mechanics*, which, like anthroposophy, seem to give pretty good results although, as Richard Feynman remarked, nobody really knows why and it's safer not to ask.

> *I think I can safely say that nobody understands quantum mechanics.*[112]

And Bohr has been repeatedly quoted as saying,

> *Anyone who is not shocked by quantum theory has not understood it.*[113]

*

For Steiner and for those of us who have experienced the validity of his observations, the worlds of soul and spirit are just as real as the physical world. A wrong thought, as he says, can be like a bullet fired into the spiritual or astral world. The popular image of Bohr's electronic atom doesn't correspond to anything in the physical world, but it actually exists in the soul world, where it carries its Ahrimanic cargo around and sprays its influence down into the etheric and physical and up into the spiritual. It is consistent with the views of Heisenberg and his colleagues to say that the physicists' atom of Steiner's time was a mental artifact, a model designed to reproduce experimental observations. It never existed in the physical world but it is nonetheless real. Having been conceived and developed in the minds of physicists, it acquired a life of its own, penetrating what is sometimes referred to as the "public mind" and becoming a very convenient means of transportation for Ahriman's ideas for the future of a humanity bound in the physical world and unconscious of the spirit. The purely electrochemical description of human physiology leaves us in the condition of electrically operated robots with Ahriman at the control panel. There is no need for me to expand on this, for as the old saying goes, "*Si monumentum requiras, perspice.*" If you require evidence of Ahriman's activities, look around.

*

The contrast between Steiner's public and private utterances recalls the struggles of human beings over the ages to find means of communication suitable to their audiences. Steiner might be accused of trimming his sails according to the prevailing winds or, like the much-maligned Averroes, of adhering to the double-truth theory, according to which contradictory statements can be simultaneously true. For several reasons these accusations would be false.

Like many great teachers, Steiner struggled with the problem of communicating esoteric knowledge to the uninitiated, using both schematic and pictorial methods to do so. The medieval Christian Church, knowing that most people would be unable to understand complex theological discussions, presented the fundamentals of Christianity by means of mystery plays. The great Islamic philosopher Averroes thought that the style and content of education should be adjusted in accordance with the capacities of those receiving it. Most people, he observed, can cope only with knowledge in the imaginative, pictorial form given in the Koran. Those who are able to follow a train of thought and to reach probable conclusions can receive instruction based on the Koran and theology, while those who seek truth in its rational essence, can use the material of the Koran for logical penetration. These ideas were condemned by the Muslim theologians and later contributed to the belief that Averroes had embraced the "double truth" theory, in spite of the fact that he specifically asserted that truth cannot contradict truth. Like Steiner, Averroes was simply following a principle which runs through teaching of almost every kind, whether esoteric or exoteric, namely that it is at best a waste of time and at worst dangerous to give people knowledge which they are not equipped to handle. Popular science writers, attempting to convey knowledge that they often don't understand to people who don't realize that they are receiving a garbled version of what real scientists think, make things easy for Ahriman. As Richard Feynman says in the introduction to his *QED*,[114] "Many 'popular' expositions of science achieve apparent simplicity only by describing something different, something considerably distorted from what they claim to be describing." I have heard garbled versions of anthroposophical knowledge that had the same effect.

To take the matter a stage further, it must be remembered that when Steiner was speaking to theosophists and anthroposophists, he wasn't addressing a room full of initiates, so although he was able to let his esoteric hair down a little more than in his public lectures, we must still take it that the pictures he gave were the nearest he could get to the spiritual reality that he experienced. The visual image that flashes into our minds of little Ahrimanic imps flying about on tiny electrical particles may be as near to the spiritual truth as the visual image Bohr made of electron orbits was to the physical truth, and may be as close to actuality as we can get in our present state of enlightenment.

*

In 1923 Steiner was horrified by the electromagnetic interpretation of light, and yet in 1904 he had told us, apparently in a matter-of-fact way, that thought is composed of electricity. It might seem safe to say that when the divine intelligence descended into the souls of human beings, it was not in the form of electricity and that our thinking has been degraded into the form desired by the beings who wish to take control of it. And yet certain passages from one of the Karma Lectures (Dornach, July 1924) indicate that it's not quite so simple.

Epilogue — 225

In Chapter VI, Section (iv) I quoted a paragraph from this lecture on the subject of the battle between Michael and Ahriman over the fate of the divine intelligence. Steiner continued with a description of a crucial event that took place behind the scenes at the beginning of the fifteenth century. "

> *In Atlantean times, when the Cosmic Intelligence had taken possession of the hearts of men, such an event had taken place; and now for the present earthly realm it once again broke forth in spiritual lightning and thunder. In the age when men were conscious of the earthly historic convulsions only, the earth appeared, to the spirits in the supersensible worlds, surrounded by mighty lightnings and thunderclaps. The Seraphim, Cherubim and Thrones were carrying over the Cosmic Intelligence into that member of man's organisation which we call the system of nerves and senses, the head-organization. Once again a great event had taken place. It does not show itself distinctly as yet, it will only do so in the course of hundreds or thousands of years; but it means that man is being utterly transformed. Formerly he was a heart-man; then he became a head-man. The Intelligence becomes his own. All the power and strength that lies in the domain of the Seraphim, Cherubim and Thrones now had to be applied in accomplishing a deed such as takes place only after many aeons. And one might say: What Michael taught to his own during that time was heralded in the earthly worlds beneath with thunder and lightning. This should be understood, for these thunders and lightnings must become enthusiasm in the hearts and minds of anthroposophists.*

So these stages of the descent of the divine intelligence were accompanied by spiritual electrical storms, for I don't believe that Steiner was being merely metaphorical when he spoke of "spiritual lightning and thunder." What I do believe is that the electricity that has become an instrument of Ahriman is an earthly descendant of the spiritual energy that came into the world with the heavenly intelligence and, as I suggested before, provides the vehicle but not necessarily the content of our thinking. Forty years ago, Marshall McCluhan told us that "the medium is the message," but, as the Gershwins would have said, "It ain't necessarily so!"

*

While we are on the subject of inconsistencies and contradictions, it is as well to recognize that the account of the descent of the divine intelligence given by Steiner in *The Driving Force of Spiritual Powers in World History* makes no mention of the Archangel Michael and leaves the transference in the hands of the Exusiai rather than the Seraphim, Cherubim and Thrones. While I don't think it's necessarily foolish to concern oneself with consistency,[115] it remains true that the same event seen from different angles is likely to generate different reports. A private on the ground in the war zone may say that his sergeant is in charge, whereas someone at HQ thinks it's the generals, and another person in Washington thinks it's the President. The point is that they are all correctly describing parts of a whole and that someone writing the story of the war might cover it in turn from each of these angles.

*

We communicate through many different media, not all of them involving words; we have the spoken word of direct conversation, accompanied by facial expressions and body language, and the spoken word by way of telephone, radio, and television; and we have the unspoken word in the form of handwriting, typewriting, or printing, not to mention texting, Facebook, and Twittering. In addition to these we have music, which some people who ought to have known better have claimed not to be a form of communication at all, and various other forms of art. It would be hard to deny that the nature of the medium imposes limitations on what can be communicated, and I am willing to go out on a limb and assert that, for good or ill, the face-to-face spoken word has the greatest potentiality for communication. Although ignorance and deliberate obfuscation keep language in a state of continuous degradation that makes it more and more difficult to express one's thoughts in a way that is both accurate and comprehensible, it is also in a state of continuous enrichment because it has to be adapted to convey new meanings. From a purely utilitarian point of view it is possible to maintain that all possible messages are implicit in the vocabulary, grammar and syntax of our language, just as it might be maintained that all possible pieces of music already exist in potential in the notes of the various scales. People still keep writing new tunes, however, and words have many shades of meaning. In both cases, music and words, the need to explore meaning, communicate new experiences or find new ways of interpreting old ones, deeply affects the evolution of the language. In saying this, we assert that meaning – the message – exists independently of language.

In a very restricted sense it is true that the spoken word is made of air, just as the written word is made of ink and the Venus de Milo is a piece of rock. In the same sense it is true that thought consists of electricity. The fact that our thinking depends on a complex system of electrical currents may impose restrictions on its content, and it may be that if our thinking about our thinking rests on the same currents, we shall never know what the restrictions are. And if you are at this moment thinking that this is where

Steiner, anthroposophy, and *The Philosophy of Freedom* come in, you are right. "The dawn of the Michael Age is at hand; hearts begin to have thoughts." One of the main goals of anthroposophical work is to free our thinking from material and sensory restrictions – in other words, to free it from the grasp of Ahriman. Heart thinking, I am fairly sure, is not made of electricity.

*

It would be hard to contest the assertion that the atom described in physics textbooks is to a large extent a mental construct, but we can't help feeling that there must be something out there that exists independently of us and our efforts to track it down. This is the age-old question of the thing-in-itself that Heisenberg and company implicitly decided was beyond the scope of atomic physics and Born would gladly have referred to the philosophers. The distinguished physicist J. A. Ratcliffe, who was my Director of Studies at Cambridge, insisted that such questions as "What does light *really* consist of?" were meaningless. The only thing that matters is what you can actually observe, and we seem to be stuck with the same problem when we ask what electricity *really* is. To the physicist the question has no meaning, but in view of Steiner's remarks on the subject, the student of anthroposophy has no option but to pursue it.

We might, in fact, adapt a comment about matter, already quoted: "When one speaks about *electricity* in the sense of a modern physicist, *electricity* is no longer *electrical*... In this case they are right..." Two centuries ago, electricity was thought of as a fluid – a kind of substance different from ordinary matter, but still in some sense substantial. A hundred years ago we thought of the electron as a bit of matter with an electric charge stuck to it, and for many practical purposes this picture is still perfectly adequate. But, as we have seen, the electron is characterized fundamentally only by a set of possible interactions – not by what it *is* but by what it *does*.

Something quite similar applies to the esoteric knowledge that Steiner gives us – it is all about capacities and interactions. It has never occurred to me to ask what the astral body is made of, or what the ego is made of, or what the Saturnian warmth was made of, or what the cosmic intelligence is made of. We do our best to develop capacities and experience interactions. You may not wish to investigate electricity by deliberately giving yourself shocks, and if you do, the shock will not necessarily generate the flash of insight that Steiner spoke of in his *1/28/23* lecture. Even so, the experience will tell you something. In my room at Cambridge I had a gas ring and an electric heater with an element made of bare wire that glowed red hot. The gas ring was connected to the copper gas pipe by a flexible tube, and the valve was at the end of the pipe, which, in the nature of things, made an excellent ground. If I ran out of matches, a fairly frequent occurrence, my technique was to switch on the heater, hold the gas ring to the red hot element with one hand and operate the valve with the other. On one occasion the gas ring touched the bare wire and my body provided an easy path for the electricity to go to ground. It was a nasty shock and, unlike Steiner, I wasn't aware of an inner moral quality; but the experience stayed with me and I recalled it forty years later when some spinal problems made it necessary for a neurologist to check my circuits. There was still no great moment of enlightenment, but I did realize in a new way the essential weirdness of electricity, the uncanny nature of the feeling produced by the shock, the momentary change of consciousness. The idea of an electric current as a flow of electrons is a useful concept and fits many experimental facts very well, but the sensation produced by an electric shock feels more as if the body is put temporarily into a different state, reminiscent of the state of strain produced by the elongation of a spring. In the old days of radio before the transistor was invented, we needed heavy batteries of 90 volts or more to make the electron tubes work, and these were known as "high tension batteries." In those far-off times, too, a spark plug connector was known as a "high tension cable," and if you didn't want anyone to steal your motorbike you simply removed it and carried it around in your pocket. (On one occasion, I forgot that the cable was in my pocket and spent an hour trying to figure out why the engine wouldn't start.) Anything connected to the poles of a battery or an induction coil is put into a state of tension analogous to that of the stretched spring. This, I think, is what happens to the nervous system, changing our relation to perception, feeling and thinking. It seems to me to be very reasonable to suppose that at the same time something happens to the connections between the members of the human organism (the physical, etheric and astral bodies and the ego) allowing penetration by spiritual influences that may be anything but benign. This, of course, is highly speculative, as any comments on the moral or immoral effects of electricity are almost bound to be unless they come from a deeper perception than most of us have.

    This still doesn't tell us what electricity is, but we shouldn't find this surprising. One of the most frustrating things about the universe is that ultimately we don't know what *anything* is. If we go in one direction, we come to spirit and perhaps recall that when we were new to anthroposophy we kept expecting Steiner to tell us what spirit is, and he

Epilogue — 231

never did. In the other direction we come to matter, which, to say it yet again, turns out to be immaterial. Geometry is about points, lines, and planes, but when we ask what these objects are, geometry has no answer – unless you consider "undefined terms" to be an answer. The universe is like that. At rock bottom there are realities that can neither be defined nor analyzed, only perceived.

*

Now I'll start again from a different point on the landscape. As I have pointed out, to say at the same time that the atom is real and that the atom is a mental construct is not necessarily a contradiction. It was only the beginning of the transference of the divine intelligence into the human intellect that made the first atomic theories possible, so that the atom could be constructed out of thought. From the time of Democritus to the time of Planck there was always opposition to the idea that atoms were actual physical particles. (It is as well to bear in mind that there are difficulties involved in explaining what "actual physicality" is; here it is just shorthand for the supposition that the atom has the same kind of existence, whatever that is, as a cannonball.) It was only in the early twentieth century that this resistance died out among professional scientists, and it was still strong enough in the Vienna of 1906 to precipitate Bolzmann's suicide.

Meanwhile the situation was changing to the extent that in the early years of the twentieth century it became close to heretical to question the proposition that the atom and all its attributes exist independently of the questing scientist. Max Born's remarks on the supposed objectivity of science bear repeating here:

In 1921 I believed – and I shared this belief with most of my contemporary physicists – that science produced an objective knowledge of the world, which is governed by deterministic laws. The scientific method seemed to me superior to other, more subjective ways of forming a picture of the world... In 1951 I believed none of these things. The border between subject and object had been blurred, deterministic laws had been replaced by statistical ones, and although physicists understood one another well enough across all national frontiers, they had contributed nothing to the better understanding of nations...[116]

Paradoxically enough, the attitude encountered and shared by Born as a young man continued into the 1930's even when Einsteinian relativity and Heisenbergian quantum physics had shown the intimate connection between the observer and the observed. It was so strong, in fact, that many eminent philosophers and scientists threw up their hands in horror when Sir Arthur Eddington set out to demonstrate that there is an essential element of subjectivity in the make-up of physical science.

Eddington (1882-1944), who was a deeply committed member of the Society of Friends, is most famous for what must be regarded as one of his lesser achievements; the verification of Einstein's prediction of the bending of light rays as they pass close to the sun. His work on the internal constitution of stars placed him in the highest rank of astrophysicists, and his quick and comprehensive understanding of Einstein's General Theory of Relativity gave him a very special place among the educators of his time.

Stated as briefly as possible, Eddington's proposition was that the universe described by physical science is a fundamentally different entity from the "objective" world of everyday experience. This is not to say that people have their own individual versions of physical science, but that the structure and content of physical science are dictated by the selective nature of its admission process.

The selection is subjective because it depends on the sensory and intellectual equipment which is our means of acquiring observational knowledge. It is to such subjectively-selected knowledge, and to the universe which it is formulated to describe, that the generalizations of physics – the so-called laws of nature – apply.[117]

This subjectivism is largely collective rather than individual, since what we allow into science is filtered by processes that we all have in common, but there is also an element of subjectivism brought about by individual decisions on the admissibility and interpretation of observational data.

In the 1890's and early 1900's, the numerical values of what were known as the "constants of nature" – such things as the mass of an electron, its electrical charge and the ratio of its mass to that of the proton – were treated as if they were brute facts. They were calculated from experimental results and there was no particular reason why they shouldn't have had different values. What Eddington did that seems to have upset people even more than the idea that the cherished laws of physics were somehow subjective, was to suggest that these values could be calculated without doing any experiments at all, just by considering the type of input and the method of calculation – in a word, epistemologically. In 1934 he made his views available to the general public in his *New Pathways in Science*, which included a calculation of the proton-electron mass ratio. This, according to him, was the ratio of the roots of a very straightforward quadratic equation obtained by considering the nature of the experimental measurements and their application – but without actually doing the experiment. By 1939 he had developed his ideas to the point of opening Chapter XI of *The Philosophy of Physical*

Science as follows:

> *I believe there are 15,747,724,136,275,002,577,605,653, 961,181,555,468,044,717,914,527,116,709,366,231,425, 076,185,631,031,296 protons in the universe and the same number of electrons.*

Eddington went on to say that he hadn't actually counted them and that if he had thought that anyone would ever count them, he wouldn't have published his calculation. (Eddington's sense of humor helped to make his books very readable, even for people with very little scientific background, and also provided a supply of banana skins for his opponents to slip on.) Electrons are indistinguishable from one another and do not have exact locations, so, as he said, they don't make "very promising material for counting." He thought that the number was inherent in the structure of quantum physics rather than in some objective "actual physical" universe. Eddington died long before anyone had ever heard of McLuhan, but if Eddington's view of physics is correct, it could reasonably be argued that McLuhan's famous dictum applies to it.

As far as Newton was concerned, when God created the universe he could have made whatever number and variety of "hard, massy particles" he thought appropriate. Eddington, however, wasn't talking about what God created but what it was possible for a modern physicist to think. But he was, at the same time, convinced of the validity of intelligent mysticism. The world of physical science was not the great, objective, all-embracing edifice that we had once taken it to be. "The purely objective sources of the objective element in our observational knowledge have already been named," he says emphatically in *The Philosophy of Physical Science*. "They are *life, consciousness, spirit.*"

It was not only that the scientific community and the philosophers on its fringes found Eddington's ideas outrageous. What may have worried them the most was that he had a gift for communication unparalleled among writers of popular scientific literature, so that his apparently subversive ideas could be communicated to the tens of thousands of people who read his books. It is unfortunate that the controversy obscured his magnificent achievements as, by general consent, the first astrophysicist and the leading authority in the English-speaking world on Einstein's General Theory of Relativity. By the time I went to Cambridge as an undergraduate in 1953, nine years after Eddington's death, most people remembered very little about him. If his name was mentioned, the usual response was, "Eddington – wasn't he the chap who said that there were 10^{79} electrons in the universe? Ha, ha, ha!"

Was Eddington's approach valid? Well, no and yes. The answers that he obtained for such physical constants as the proton-electron mass ratio and the fine structure constant were very close but not quite close enough – at least according to the generally accepted values – and quantum physics was still in its fairly early stages when he was working out the epistemology described in *The Philosophy of Physical Science* (1939). As he remarks on page 51, "The terminology of quantum theory is now in such utter confusion that it is well-nigh impossible to make clear statements in it." His attempt at a fundamental theory was unfinished at the time of his death in 1944 and even the most sympathetic readers have found it problematical. And yet there are still people, myself included, who find his work inspirational and feel that he was on the right track. In one very important respect he has been thoroughly vindicated; one of the major preoccupations of present-day physicists is the search for a theory that will account not only for the masses, charges, and other characteristics of the proton, electron, and all the other particles that have been either discovered or predicted, but also for the laws governing their interactions, including gravitation. No one believes that these items are brute facts – as Parmenides would have said, "Nothing can be which cannot be thought" – so there must be a fundamental, rationally expressible framework into which they all fit. This is the "theory of everything" that continues to elude physicists, and their continued search for it indicates the strength of their urge to find a thought structure that generates the structures of their physical world. To put the matter another way, they are using the substance of thinking to create the fundamental entities of their physical world. This endeavor certainly sheds a new light on Steiner's contention that thought and atom are made of the same thing.

I say "their" physical world because, as Steiner and Eddington pointed out in their different ways, the world of physical science differs radically from the world of everyday experience. Once again, here is Steiner speaking in January 1923:

> We have to become clear about what we actually do when, in our thinking, we cast inwardly experienced mechanics and physics into external space. That is what we are doing when we say: The nature of what is out there in space is of no concern to me; I observe only what can be measured and expressed in mechanical formulas, and I leave aside everything that is not mechanical.

And here is Eddington, "speaking as anyone might do who depends not on specialized knowledge but on that which is the common inheritance of human thought."

We recognize that the type of knowledge after which physics is striving is much too narrow and specialized to constitute a complete understanding of the environment of the human spirit. A great many aspects of our ordinary life take us outside the outlook of physics... Any discussion as to whether they are compatible with the truth revealed by physics is purely academic; for whatever the outcome of the discussion, we are not likely to sacrifice them, knowing as we do from the outset that that the nature of Man would be incomplete without such outlets...[118]

What distinguishes us most clearly from the rest of creation and unites our religious, artistic and scientific faculties is the impulse to attain "something after which the human spirit is bound to strive," namely truth.

Clearly, as Eddington says, there is "something to which the truth matters" and it belongs "in our own nature, or through the contact of our consciousness with a nature transcending ours, [with] other things that claim the same kind of recognition – a sense of beauty, of morality, and finally at the root of all existence an experience which we describe as the presence of God."

Steiner and Eddington, out of very different life experiences, agree in asserting that physical science seeks a kind of truth that excludes all the *human* qualities that are most precious to those whose devotion to the truth embraces beauty and goodness. In *New Pathways*, Eddington illustrates this with a reference to Dante Gabriel Rossetti's Blessed Damozel, who looks down from the "gold bar of Heaven" and sees the earth spinning "like a fretful midge."[119]

Looking from the abode of truth, perfect truth alone can enter her mind. She must see the earth as it really is – like a whirling insect. But now let us try her with something fairly

> modern. In Einstein's theory, the earth is a curvature of space-time... What is the Blessed Damozel to make of that?

If, like Einstein, she sees the earth as a curvature in space-time with a spin that is the ratio of two components of curvature,

> she will be seeing truly – I can feel little doubt of that – but she will be missing the point. It is as though we took her to an art gallery, and she saw ten square yards of yellow paint, five of crimson, and so on.

There have been many physicists with deeply spiritual aspirations, but I have found Eddington to be by far the most helpful in providing a tenable view of physical science that illuminates its specialized relationship to the whole human being.

*

In his scientific courses, Steiner was usually quite polite about modern science, but he put things more bluntly in the Karma Lectures:

> The blundering, inadequate, and frequently repulsive attempts of modern natural science must be transmuted by a spiritual world-conception, so that a true reading of the Book of Nature will arise from them. This is the impulse of Michael. Now that the Intelligence administered by him has come down to us, it is his impulse to lead us again to the point where we shall read once more in the Book of Nature. In reality, all of us in the Anthroposophical Movement should feel that we can only understand our karma when we know that we are called to read once more, spiritually, in the Book of Nature — to find the spiritual background of Nature.

Steiner made these remarks only a few years after he had confirmed the validity of his earlier work on Goethe's world conception, so we have a very good idea of what he meant when he spoke of reading spiritually in the Book of Nature.

We should also bear in mind that while Steiner was delivering the Karma Lectures, Bohr, Heisenberg, Pauli, Born, and Schrödinger were creating refined and elegant methodologies that would revolutionize atomic physics and meet with considerable success in dealing with its problems. Steiner, however, was talking about science in general, not just about atomic physics. Many elegant scientific theories have been produced in the past several centuries, and most of them have turned out to be inadequate. Whether or not everything is made of actual, physical atoms, the atoms that we find in physics texts are clearly products of the way we think and the way we base our experiments on our thinking. That being the case, it's not surprising that we can fit them into elegant logical structures. The chipmunk is an elegant creature but not a product of our thinking; as far as I know, no one has made an elegant theory of chipmunks.

Natural science as referred to in this last quotation includes physics, chemistry and the biological sciences. At one time I thought I could observe in Steiner's opinions of these sciences a continuum of the increasing perception of blunderment, inadequacy and repulsiveness as we pass from the more physical to the more biological. He was deeply interested in modern physics and wanted its experimental results, if not its theories, to be part of the Waldorf Schools curriculum, specifying in particular the knowledge of X-rays and alpha, beta and gamma rays. Most disturbing to him were the penetration of electricity into atomic physics and the impulse to explain human inner experience in terms of the external physiology of the senses; but we have to live with electricity and in some sense wrest it from Ahriman's hands, while the Ahrimanic image of the human being is something that has to be resisted at all times and at any cost.

In 1904, when Steiner spoke of the atom as coagulated electricity, the physicists were blundering around with inadequate atomic models, but from 1913 onwards the Bohr atom became a thing to be reckoned with. The atomic physicists might "always put you in the wrong," but their efforts were not despicable. After 1925 the immateriality of matter became more evident, and quantum physics, applied pragmatically because no one really understood it, provided a reliable means of quantifying atomic processes. For everyday purposes, however, it was still advisable to take the materiality of matter into consideration — for instance, when buying potatoes or contemplating the effects of jumping off a cliff. This is another way of saying that the gap between the fundamentals of atomic physics and the fundamentals of everyday life had grown much wider.

Is there any serious hope or fear that the fundamentals of everyday life will one day be predictable from the fundamentals of atomic physics? In my view, the answer is an obvious "No," and perhaps the least important reason for this is that fundamental physics is still incomplete and subject to serious disagreement about how it should proceed. A deeper consideration is that as long as we experience inputs into our life from a spiritual source that has nothing to do with the diligent strivings of well-meaning physicists, we know that the physical world is a descendent of the void from which God created the Heavens and the Earth, or the clay from which Adam was formed, or the warmth from which the hierarchies formed the human race. It is there to receive the impress of the spirit.

*

More recently, in re-reading the First Scientific Lecture Course[120], given to the teachers at the first Waldorf School in 1919, I was forced by the following passage from Lecture 9 to reconsider my view of a continuum of blunderment.

Particularly in physics, quite a few concepts have proved to be extraordinarily fragile, and this fact has a much stronger connection than we might think with the misery of our times. Isn't it true that when people think in sociological terms, we notice right away how their thinking has gone awry? Most people don't even notice that, but we are able to notice it because sociological ideas have an effect on the human social order. But we don't really form any satisfactory idea of how deeply the concepts of physical science affect all of human life, so we are ignorant of the damage that has, in truth, been caused by the often horrific ideas of modern physics.

Obviously this is something that I have been at great pains to bring out, but without using such an extreme qualifier as "horrific." Steiner mentions the particular case of a German professor of physics who considered that the course of the 1914-18 war had been deeply affected by the lack of an adequate connection between the university science labs and the military and felt that in the future there should be a link straight from the institutes of experimental science to the general staffs.

"The human race," Steiner continues, "must change its ideas, and must change them in many areas. If we can decide to change them in such an area as physics, it will be easier for us to do so in other areas, too. Those physicists who go on thinking in the old way, however, won't ever be far removed from this nice little coalition between the scientific institutes and the general staffs."

In order to avoid possible confusion, I must remark that the "new way" implied by this last sentence is not quantum physics but Goethean physics.

The pipeline from the lab to the battlefield is a manifestation of what we call "patriotism" if we approve of it and "nationalist imperialism" if we don't. The impulse to defend one's country is not without merit, but desire to take over someone else's is another matter. We can, however, regard this as symbolic of another pipeline, which has the potential to be more insidious and even more far-reaching in its effects. This is the pipeline from the mind of the physicist to the soul of the student or general reader, where the battle with Ahriman is taking place.

Reading the works of the scientists involved in the evolution of atomic physics, one usually has the impression that they are deeply preoccupied with the effort to understand and explain the phenomena they are working with – the properties of gases, the discharge tube, spectra, cloud-chamber pictures, collisions in high energy accelerators and so on – rather than with the explanation of the whole universe.

> *In agreement with Heisenberg, Bohr emphasized that the aim of physics was to predict and coordinate experimental results, not to discover the reality behind the phenomenal world. 'In our description of nature', he wrote in 1929, 'the purpose is not to disclose the real essence of the phenomena, but only to track down, as far as is possible, the relations between the manifold aspects of our experience.'*[121]

It would be hard to endorse Steiner's view of the status of quantitative science more comprehensively or to express more trenchantly the view that our mental images of particles and waves are merely visual aids.

Some physicists, however, take time off from their professional work to speculate on economics, morals, history, philosophy and religion and join their less well-informed and less scrupulous journalistic colleagues in giving us the feeling that the ultimate aim of atomic science is to show that everything, including human consciousness, is predictable, at least in principle, from the properties of atomic or subatomic particles. If this ever came to pass, it would become plain that soul and spirit, selfhood and consciousness are all illusions, a consummation greatly to be desired by the spiritual beings who wish to deny us our inner freedom. *It must be noted that for these beings to achieve their desire, it is not necessary that all or any of these conclusions should be true, but only that most people should believe them.*

A glance through the popular scientific literature of the present time shows that there is reason to believe that this may become the case within the foreseeable future. The increasing attention paid to studies of the human nervous system, the chemical mechanisms associated with consciousness, and self-referential systems capable of simulating some kind of self-awareness, has given rise to a genre in which complex and profound questions are simplified to the point where the unsuspecting reader can be given the illusion that he understands them, while the dilemmas of twentieth century quantum physicists conveniently remain unmentioned. But modern psychology depends on physiology, physiology depends on biology, biology depends on chemistry, and chemistry depends on physics; and, deep down, the wonderful thing is that nobody understands physics.

Long ago, as Steiner tells us, the church fathers abolished the spirit. In the twentieth century some of the scientific and pseudo-scientific fathers went a long way towards the banishment of the soul and the invalidation of consciousness itself. We need to understand how our feelings about ourselves and our fellow human beings would be affected if we believed that all our inner processes were merely the results of events on an atomic level, which may be considered to proceed without regard to our volition.

Suppose that all the time we thought that we were making decisions it was really just our physical bodies making them for us. If that is the case we might ask, "Who am I?" or just, "Am I?" What is there that can still be said for the old-fashioned idea of a conscious self, giving some direction to the mental and physical traffic? We experience a degree of independence in our thought processes and a sense of continuity, not only in our own personal history but also in the lives and experiences of the people around us, that argue vehemently against the idea that consciousness has emerged from a sequence of senseless physical processes.

As a Waldorf teacher I have spent most of my time in a community in which the objective existence of soul and spirit is taken for granted and in which current efforts to account for consciousness in purely material terms are barely given the time of day. For many scientists who think about such things, the position is precisely the other way around. At this time in history neither of these positions is tenable for people who wish to be responsible for their own destinies. We believe that our life decisions should be based on conviction rather than convenience, but a conviction rooted in ignorance and maintained by turning a deaf ear is meaningless. How Ahriman must chuckle when he hears ignorant people speaking abusively about science and technology! The great thing is, however, that ignorance is one human ill that can actually be cured. Knowledge may or may not be power, but without it we shall never attain the degree of alertness which Steiner tells us is essential if Ahriman's quest to bind humanity permanently within the material world is not to succeed. So take courage. The strength and inevitability of the Ahrimanic use of atomic science may seem overwhelming, but anthroposophists are not supposed to be frightened of dragons; as Bilbo's grandfather used to say, "Every worm has its weak spot."[122] The first thing to do is to get to know your worm.

Appendix:
A Brief Note about String Theory

As I've mentioned, when I first became interested in science it seemed that the actual sizes and electrical charges of the fundamental particles of matter – the proton and the electron – were brute facts for which no explanation was required or available. This was in the mid-1940s and my pre-war textbooks were a little old-fashioned, so it was some time before I learned much about neutrons, neutrinos, and the multitudes of "funny particles" with all kinds of different masses that were being discovered. Some further time elapsed before I received the shocking news that gravitational and electromagnetic forces had been joined by a strong force and a weak force that nobody seemed to understand. The final straw, as I then thought, came long after I had graduated from Cambridge, when we were informed that protons and neutrons were not fundamental particles after all, but composed of quark triads, and that the forces that held them together or pulled them apart acted through the agency of exchange particles. It seemed to me that there had been something very calm and orderly about the schoolboy image of the Rutherford-Bohr atom with its apparent resemblance to the solar system. I thought there were now far too many particles, and I particularly disliked exchange particles. Unlike Newton, I had never had any difficulty in imagining action at a distance – the sun's gravity, for example, acting on the earth with no material intermediate agency. According to Newton, gravitation shows the presence of God, and, according to my understanding, God is present fully everywhere and locally nowhere, so I couldn't imagine what the problem was. Unfortunately I was in a minority of one.

So now we had four forces; the gravitational and the electromagnetic, respectable, well-known forces that conformed to common-sense mathematical rules and had been around for donkeys' years; and the nuclear forces, strong and weak, newcomers which did not fit at all comfortably into my picture of the way the world ought to be. Everyone else seemed to like them, however, so I had to put up with it. We also had a multitude of particles – large ones, like protons and neutrons, medium ones known as mesons, little ones, like electrons and positrons, and infinitesimal ones like neutrinos. What the physicists wanted more than anything was something known as a Unified-Field Theory, which would unite all the forces in one set of equations and give some sort of rhyme and reason for the existence and vital statistics of all the particles. No such theory has yet been found, although some less far-reaching unifications have been achieved.

The sensational thing about String Theory was that when it first appeared, it gave the promise of generating exactly this longed-for result. Put very briefly, it proposes that the world-structure is built on many more dimensions and contains many more particles than we have yet observed, and that all these particles arise from the application of simple and beautiful laws to the vibrations of a string. In the early days of the theory, it was hard for its proponents to gain a foothold in the academic world where most research is done, but soon it swept the board, and more recently it has been almost impossible for a non-stringy physicist to get a university position. This has happened in spite of the fact that string theory has failed to live up to its original promise.

One of the legitimate requirements for a scientific theory is that it should make predictions that can be tested experimentally. To put it very simply, science takes a step forward when such a prediction is verified or falsified. Such an event occurred when Louis de Broglie, having learnt about Einstein's theory that light energy travels in particle-like packages, made the inverse hypothesis that a stream of particles, such as electrons, should have wave-like characteristics. De Broglie calculated the wavelength of the associated waves, and in 1927, three years after the publication of his work, the first electron diffraction experiments verified his predictions. The thing that increasingly troubles the present generation of physicists is that string theory, "beautiful" and "elegant" as it may be, has been with us for thirty years and has not succeeded in predicting anything new of major significance that can be verified experimentally. In the century that between 1900 and 1970 produced the General Theory of Relativity, the Rutherford-Bohr Atom, the Quantum Theory, Wave Mechanics, Quantum Electrodynamics, and Quantum Chromodynamics, all of which made verifiable predictions, brought great successes and posed intractable problems, this was a deeply anomalous situation. The awful thought has, in fact, crept into the string theory community that, contra Keats, no matter how beautiful the theory may be, it may not be true.

String theorists are, however, extremely resilient, and there is strong opposition to this perception. As Lee Smolin puts it in his *The Trouble with Physics*[123],

> It seems to me that any fair-minded person not irrationally committed to a belief in string theory would see this situation clearly. A theory has failed to make any predictions by which

it can be tested, and some of its proponents, rather than admitting that, are seeking leave to change the rules, so that their theory will not need to pass the usual tests we impose on scientific ideas.

What struck me as slightly comical about this situation was that the string theorists seem to want to do exactly what modern scientists and historians have often reproached their ancient forbears for doing. The contemplative armchair has been replaced by the laptop computer, but the principle is the same. The philosophy, the world picture, the theory must be true to inner vision, and if the external world appears to disagree, there must be something wrong either with it or with our perception of it. This was the view of Parmenides and Melissus twenty-five centuries ago, and it was controversial even then, when inner vision was much more alive than it is now.

Even more striking is the admission that string theorists don't actually know what string theory is. Brian Greene, the most prominent popularizer of the theory, says in his latest book, *The Fabric of the Cosmos*, "More than three decades after its initial articulation, most string practitioners believe we still don't have a comprehensive answer to the rudimentary question, What is string theory?" or as Nobel Laureate David Gross remarked at the end of a string theory conference, "We don't know what we're talking about..."[124]

Beauty is indefinable and the poet greatly oversimplifies its relation to truth – unless you wish to define beauty in terms of truth, in which case the last line of the famous ode becomes a tautology. One element of beauty may be the truthfulness through which an object obeys its own inherent laws; but these laws, whether the object is a painting, a sonata, or a theory, are not necessarily the laws of the universe. The string theorists have made some progress in creating a universe, but it may not be the one that we actually live in.

Endnotes

1   Born, *Physics in my Generation*, Springer-Verlag, New York, 1969.

2   This discussion is greatly indebted to *Early Greek Philosophy*, Penguin 1987, translated, edited and introduced by Jonathan Barnes, and, of course, to Rudolf Steiner, who provided the spiritual key to the working of the ancient Greek mind.

3   Jonathan Barnes, *op. cit.*

4   Originally part of the cycle, *The Driving Force of Spiritual Powers in World History*, Steiner Book Centre, North Vancouver, 1972. Reprinted in *The Inner Nature of Music and the Experience of Tone*, Spring Valley, NY, 1983.

5   Simplicius (sixth century AD) was a Neoplatonist philosopher from Cilicia (modern Turkey). His writings include quotations from, and commentaries on, many ancient sources that are now lost. He was one of the philosophers banished from Athens by Justinian in 529 AD.

6   See Averroes: *On the Harmony of Religion and Philosophy*. Tr. and ed. G. F. Hourani; Luzac, London 1976

7   Editorial addition by Jonathan Barnes, *op. cit.*

8   Barnes, *op. cit.*

9   Obviously Melissus had never heard of quantum fluctuation, the process which is supposed to be a possible source of material for the big bang.

10   Rudolf Steiner, *The Riddles of Philosophy*, Anthroposophic Press, Spring Valley, New York, 1973.

11   J. B. S. Haldane (1892-1964) was an eminent evolutionary geneticist who fought in the Great War and in 1923 proposed a system of hydrogen-generating windmills, the first proposal for a hydrogen-based renewable energy economy. He was a pioneer of the use of quantitative methods in biology and one of the founders of the mathematical theory of population genetics. His book, *The Causes of Evolution*, 1932, played a major part in reestablishing natural selection as the main mechanism of evolution by explaining it in terms of the mathematical consequences of Mendelian genetics.

12   McMillan, New York, 1947.

13   Even mere physical contact presents a problem for unchangeable particles. When two billiard balls collide they are both compressed slightly at the point of impact and the subsequent expansion drives them apart. It is easy enough to say that if the balls were unchangeable the duration of the collision would be zero and the force between them infinite but it is hard to attach any clearly definable meaning to such a statement.

14   This seems inconsistent with Newton's rejection of the idea of action at a distance. A discussion of that hot topic can be found in my *From Abdera to Copenhagen*.

15   Isaac Newton, *Opticks*, Edition of 1730, Dover, NY, 1952; Question 31, p.400.

16   To a physicist, *"fluid"* means "anything that flows," be it gas, vapor or liquid.

17   Boyle's Law states that the pressure of a given mass of gas is inversely proportional to the volume, as long as the temperature is constant.

Endnotes — 253

18   For every known substance that can exist in the gaseous state there is a certain temperature, known as the critical temperature, above which the substance can exist only in the gaseous state, no matter how much it is compressed. The substance can exist in the gaseous state below its critical temperature, but increasing compression eventually results in a discontinuous decrease in volume as the substance condenses into a liquid. Substances in the gaseous state below their critical temperatures are known as vapors. The critical temperature for water is 380°C, for oxygen, -183°C, and for helium, -272°C.

19   Antoine Laurent Lavoisier (1743-1794) was a professional scientist with a strong background in mathematics, astronomy, botany and chemistry. He published papers on widely varied subjects and helped in the preparation of a mineralogical atlas of France. By 1775 Lavoisier was engaged in his own chemical researches. These continued alongside his major civic undertakings until his execution in 1794. He worked tirelessly to improve agriculture and to ameliorate the lot of the common people by means of savings banks, insurance societies, and public works projects. He secured funds for those who were threatened with starvation in the famine of 1788, and in 1791 provided the treasury with an extraordinarily efficient accounting system. All these labors and the commitee work which he undertook in 1790 on coinage, public health, and the institution of the metric system, were insufficient to keep him from the guillotine. "We have no need of savants," remarked Marat.

20   *The Feynman Lectures on Physics*, Vol. 1, §4-1, Addison-Wesley, 1963.

21   Rudolf Steiner, *Autobiography*, SteinerBooks, Great Barrington, MA, 2006.

22   Rudolf Steiner, *The Boundaries of Natural Science*, Dornach, 1920; Anthroposophic Press, Spring Valley, NY, 1982.

23   Steiner, *Goethe the Scientist*, Tr. Olin D. Wannamaker, Anthroposophical Press, New York, 1950, p. 205.

24   *Ibid.* p. 254

25   In those days the word was still usually spelt *"radicle"* when used as a piece of chemical terminology.

26   Sodium, potassium, barium, strontium, calcium and magnesium.

27   Given to the teachers at the first Waldorf School in Stuttgart, 1919-20, and now published under the title *The Light Course*, translated by Raoul Cansino; Anthroposophic Press, Great Barrington, MA, 2001.

28   Encyc. Brit, 1942

29   The term "electron" seems to have been coined by G. Johnston Stoney and originally meant the unit charge on an ion in an aqueous solution. E. N. da C. Andrade refers to Stoney as "the Irishman," and gives the name as Johnstone and the date as 1894. (*Rutherford and the Nature of the Atom*, Doubleday Anchor, 1964) Richtmyer and Kennard (*Introduction to Modern Physics*, McGraw-Hill, 1947) agree with Andrade about the spelling but say 1891. J. J. Thomson in the 1940 Enc. Brit. gives 1881 and drops the *e*. Mellor (*Modern Inorganic Chemistry*, 1939) states that Stoney calculated the unit charge ionic charge in 1874.

30   Lecture to Royal Institution, 1897.

31   "The Work of Secret Societies in the World - The Atom as Coagulated Electricity"; Berlin, Dec. 23, 1904.

32   *The Feynman Lectures on Physics*, Vol. 1, §4-1, Addison-Wesley, 1963.

Endnotes — 255

33 Yes, I am aware that light is not visible.

34 John William Strutt, Lord Rayleigh (1842-1919), became very well known for his work in acoustics, optics, hydrodynamics, and electromagnetism, and for initiating the research that resulted in the discovery of the inert gases.

35 John Tyndall (1820-1893), an English physicist born in Ireland, succeeded Faraday as Superintendent of the Royal Institution in 1860, and became his great friend. Although he was a gifted poet and a very distinguished scientist in his own right, Tyndall is probably best known for his memoir, *Faraday as a Discoverer*.

36 Rudolf Steiner, *The Philosophy of Freedom*, trans. by Michael Wilson, Rudolf Steiner Press, London, 1964.

37 Like *"visible light," "white light"* is a phrase that we use in order to avoid circumlocution.

38 Cambridge University Press, 1934; Ann Arbor, MI, 1959.

39 Wien: Nobel Prize acceptance speech, 1911.

40 *Ibid*.

41 *"Action,"* in this sense, is a physical quantity for which I can think of no intuitive explanation. Dimensionally it is equivalent to energy multiplied by time, so when Planck's quantum of action is multiplied by frequency, the result has the dimensions of energy. The concept is important because of the Principle of Least Action, which provides a way of calculating the trajectory of a particle.

42 Max Planck, Nobel Prize acceptance lecture, 1920.

43   Isaac Newton, *Principia*, Book 2, Proposition XXIII, Theorem XVIII.

44   J. R. R. Tolkien, *The Fellowship of the Ring*, p.18. Houghton Mifflin, Boston, 14th printing, 1964.

45   Students of elementary calculus are familiar with derivatives. The derivative of distance with respect to time is speed and the second derivative is acceleration.

46   At the time when he was struggling with this problem, the most well-known universal constant was probably the number G that determines the gravitational attraction between two masses.

47   In current usage *quantum* refers to the product hv. Planck's phrase *the quantum of action* simply means h.

48   Helge Kragh, *Quantum Generations*, Princeton University Press, 1999.

49   "*Classical physics*" means the physics that existed up to the close of the nineteenth century and accepted the validity of Newtonian mechanics and nineteenth century electrodynamics.

50   Gino Segrè, *Faust in Copenhagen*; Viking Penguin, 2007.

51   E. A. Karl Stockmeyer, *Rudolf Steiner's Curriculum for Waldorf Schools*, Steiner Schools Fellowship, 1969.

52   Rudolf Steiner, *Practical Advice to Teachers*, trans. by Astrid Schmitt-Stegmann; Anthroposophic Press, Great Barrington MA, 2000.

53   Rudolf Steiner, *The Boundaries of Natural Science*, Dornach, 1920; Anthroposophic Press, Spring Valley, NY, 1982.

Endnotes — 257

54   This and the quotations that follow are drawn from Rudolf Steiner, *Anthroposophy and Science*, Stuttgart 1921; Mercury Press, Spring Valley, NY, 1991.

55   Steiner, *The Origins of Natural Science*, Lecture II, Dornach, Christmas Eve, 1922. Anthroposophic Press, Spring Valley, NY, 1985.

56   The use of *"philosophy"* to mean what we now call *"physical science"* persisted into Steiner's time.

57   Steiner, *The Boundaries of Natural Science*.

58   Mercury Press, Spring Valley, NY, 1985.

59   See Steiner, *Knowledge of the Higher Worlds and its Attainment*, Anthroposophic Press, Spring Valley, NY, 1983.

60   Chaucer, from "House of Fame," quoted by C.S. Lewis in *The Discarded Image*, C.U.P., 1961. Lewis was quite sure that although medieval thinkers attributed life and intelligence to stars and planets, they did not believe "that what we now call inanimate objects were sentient and purposive." My reference is, however, to put it crudely, to what people thought when they weren't thinking – their intuitive grasp of reality before it became "sicklied o'er with the pale cast of thought." They did not have to believe that the apple deliberately falls to the ground; only that the operation of the whole universe is maintained by the unimaginable hand and mind of God.

61   Francis Bacon: *The Great Instauration; Plan of the Work*.

62   More recently published as *Nature's Open Secret: Introductions to Goethe's Scientific Writings*, Anthroposophic Press, Great Barrington, MA, 2000.

63   Steiner, *Goethe the Scientist*, p. 205.

64   Quoted by Steiner, *Ibid*, from remarks by Wilhelm Oswald.

65   *Ibid*. p. 245.

66   Steiner, *The Philosophy of Freedom*, translated by Michael Wilson, Rudolf Steiner Press, London, 1964.

67   Printed in John Tyndall, *New Fragments*, D. Appleton and Co. New York, 1896.

68   Henry Thomas Buckle (1821-1862) was the author of an unfinished History of Civilization. His delicate health prevented him from obtaining much formal training but he received a high degree of education privately. After his father's death in 1840, he directed all his reading to the preparation of some great historical work. Over the next seventeen years, he is said to have spent ten hours a day on it. The first volume, which appeared in June 1857, made its author a literary and social celebrity.

69   Steiner, *Goethe the Scientist*, Section VIII.

70   *Ibid*. Section II

71   *Ibid*. Section IV

72   *Ibid*.

73   This sounds as though the scattering theory depends on the assumption that the colors are already present in sunlight. It is, however, just a convenient way of talking.

74   Dalton took $H = 1$ as his standard.

75   Gallium and indium, discovered in 1875 and 1863, respectively, react with water to form hydroxides which have both basic and acidic properties.

Endnotes — 259

76   Various sources give the number of children as anything from thirteen to seventeen.

77   There is some confusion about the discovery of gadolinium. This is the version given in Mellor's *Modern Inorganic Chemistry*.

78   "Lutecium" is now usually spelt "lutetium," but, as is my habit, when I prefer the older spelling I continue to use it.

79   Cerium also forms a stable oxide and salts in which it shows a valence of 4, creating a considerable temptation to put it in Group IV. That, of course, would leave the problem of what to do with the next thirteen rare earth elements.

80   Twelve of them were known before 1880.

81   The only exception that springs to mind is the chemical concept of valence.

82   Crookes believed that there was a link to extra-sensory phenomena and spent a great deal of time working with the Society for Psychical Research.

83   Even the best vacuum contains something like 1020 molecules per cubic centimeter. A molecule bouncing off the warm side of the vane would receive extra momentum and thereby exert a greater force on the vane.

84   There seems to be an element of mystery or confusion over the sequence of events leading to Becquerel's discovery. I have consulted several texts and found the date given variously as 1895, 1896, and 1897. Feynman gives it as 1898, but I suspect that this is through not bothering to check – a very minor infraction. My high school physics master, Hubert Siggee, maintained that the discovery was completely fortuitous, having been caused by the storage of the uranium compound near a drawer of photographic materials. Fred Hoyle tells the same story (*Home*

260 — *Rudolf Steiner and the Atom*

Is Where The Wind Blows, Mill Valley, CA, 1994). The connection with X-rays is brought out by Richtmyer and Kennard, *Op. cit.*, and various other texts originating in the 1930's, giving February, 1896 as the date of Becquerel's original announcement. The 1940 edition of the *Encyclopaedia Brittanica*, which is noteworthy for the excellence of its articles on science and on many other subjects, tells the latter version of the story in some detail, confirms that the date was 1896, and goes on to describe the work of the Curies, Ernest Rutherford and several other distinguished scientists in isolating the radioactive elements and analysing the radiations. On reaching the end of the article, I was delighted to find that the author's initials are E. Ru. – Ernest Rutherford himself! A little later I discovered that the article on the conduction of electricity had been written by J. J. Th. – J. J. Thomson – and the one on Atomic Physics by N. B. – Niels Bohr. This is like finding an article on Elizabethan drama signed by W. Sh. and one on spiritual science signed by R. St.

85   Quoted by Andrade, *op. cit.*

86   *Mass* is a more difficult concept than most textbooks would have you believe. The old-fashioned definition – *quantity of matter* – is fatally flawed but it will have to do for now.

87   Newton solved the problem of motion under central force. A particle of mass m, moving with uniform speed v around a circle, radius r, has an acceleration v^2/r towards the center of the circle. According to the Second Law of Motion this means that there must be a force mv^2/r pulling the particle towards the center. If the force is governed by an inverse square law, its magnitude is k/r^2, where k is constant. So the circular orbit is governed by the equation $k/r^2 = mv^2/r$, i.e. $mv^2 = k/r$. The value of k can easily be calculated for orbits in gravitational and electric fields, so the energy of a planet or a particle can be found for any given radius. A little further calculation shows that the square of the period of revolution is proportional to the cube of the radius, which is in accordance with Kepler's Second Law of Planetary

Motion. The calculation can easily be extended to include elliptical orbits.

88   John van Vleck, quoted by Kragh, *op. cit.*

89   Quoted by Segré, *op. cit.*

90   Poincaré (1854-1912) was an outstandingly gifted and highly influential mathematician-physicist-philosopher.

91   *Mystery Knowledge and Mystery Centres.* Rudolf Steiner Press, London, 1997.

92   Johann Paul Friedrich Richter (1763-1825) German romantic novelist, quoted by Sir Donald Tovey, *Essays in Musical Analysis;* OUP, 1935. Tovey's memory was exceedingly capacious, but his retrieval system was not infallible. As he wrote in the forward to the second impression of his essays, "I have already disclaimed for these essays any research except the verifying of quotations where they are not best left unverified."

93   To get an idea of what college physics was like in America in the 1950's, it is instructive to take a look at such texts as *Modern College Physics* by Harvey E. White (Van Nostrand, third edition, 1956).

94   In the *Supplementary Course*, Stuttgart, 1921; now published under the title, *Education for Adolescents*, translated by Carl Hoffman, Steiner Books, Great Barrington, MA, 1996.

95   St. Mark: 8;36.

96   I apologize for the physicists' habit of using *"light"* as a generic term for *"electromagnetic waves."*

97   Louis-Victor Pierre Raymond, Prince de Broglie, (1892-1987).

98 Davisson had actually written to Born in 1925, describing some of his results. Born put 2 and 2 together and asked W. Elsasser to investigate. Elsasser's results were published in 1925.

99 This is the name by which the principle is usually known; a more appropriate name is the Principle of Indeterminacy.

100 French mathematician and astronomer Pierre Simon, Marquis de Laplace (1749-1827), made extensive studies of the solar system in terms of Newton's Law of Gravitation.

101 In this connection, it's worth noting that Max Born's essay, *Is Classical Mechanics in fact Deterministic?*, shows that the whole idea of classical determinism was based on a fallacy long before quantum physics appeared. This essay, first published in 1955, and reprinted in *Physics in my Generation*, Springer-Verlag, New York, 1969, anticipates one of the most important ideas of chaos theory.

102 *Atomic Physics and the Description of Nature*, 1934.

103 Born, *Physics in my Generation*, Springer-Verlag, New York, 1969.

104 Segrè, *op. cit.*

105 Quoted by Segrè. Arnold Sommerfeld was the leading exponent of Bohr's theories in their more rigorously mathematical form.

106 Heisenberg, *loc. cit.*

107 Kragh, *op. cit.*

108 See appendix on string theory.

109   Maurice Maeterlinck (1862-1949) was a Belgian author and Nobel Prize winner whose works conveyed a sense of mystery and impending doom. Thanks to Debussy, he is probably best known for his drama *Pelléas et Mélisande*.

110   Presumably Georg Christof Lichtenberg (1742-1799), who was a great one for aphorisms.

111   At this point it would perhaps be a good idea to reread Chapter I, Section (xi).

112   Richard Feynman, *The Character of Physical Law*, Random House, 1994.

113   In *Meeting the Universe Halfway*, 2007, p. 254, Karen Michelle Barad gives a footnote citing *The Philosophical Writings of Niels Bohr*, 1998.

114   Princeton University Press, 1985.

115   Emerson: "*A foolish consistency is the hobgoblin of small minds.*" People who quote this remark often omit the word "*foolish.*"

116   For the full quotation, see the Introduction.

117   Sir Arthur Eddington, *The Philosophy of Physical Science*, C. U. P., 1939, Ann Arbor, MI, 1959.

118   Eddington: *Ibid.*

119   D. G. Rossetti (1828-1882), an English poet and painter, was a founding member of the Pre-Raphaelite Brotherhood, formed in protest against the materialism of industrialized E n g l a n d . Like Maeterlinck, he owes part of his fame to Debussy, who set *The Blessed Damozel* (*La Damoiselle Elue*) for chorus and orchestra in 1893, the year in which he started work on *Pelléas et Mélisande*.

120 Now published under the title *The Light Course*, translated by Raoul Cansino; Anthroposophic Press, Great Barrington, MA, 2001.

121 Kragh, *op. cit.*

122 J. R. R. Tolkien, *The Hobbit*, Houghton Mifflin, Boston. Bilbo noted that his grandfather's remark was not made from personal experience.

123 Houghton-Mifflin, New York, 2007

124 Quoted by Lee Smolin, *ibid.*

Bibliography

Andrade, E. N. da C.: *Rutherford and the Nature of the Atom*, Doubleday Anchor, 1964.

Averroes: *On the Harmony of Religion and Philosophy*, trans. and ed. G. F. Hourani; Luzac, London, 1976.

Bacon, Francis: *Novum Organum*, tr. Urbach and Gibson; Open Court, Chicago, 1994.

Barnes, Jonathan: *Early Greek Philosophy*, Penguin 1987.

Born, Max: *Physics in my Generation*, Springer-Verlag, New York, 1969.
Atomic Physics, Dover, New York, 1989

Eddington, Sir Arthur: *New Pathways in Science*, CUP 1934; Ann Arbor 1959.
The Philosophy of Physical Science, CUP 1939: Ann Arbor, 1958.

Feynman, Richard: *The Feynman Lectures on Physics*, Addison-Wesley, 1963.
The Character of Physical Law, Random House, 1994.
QED, Princeton University Press, 1985.

Kilmister, C. W.: *Sir Arthur Eddington*, Pergamon Press, Oxford, 1966.

Kragh, Helge: *Quantum Generations*, Princeton University Press, 1999.

Lewis, C. S.: *Miracles*, Macmillan, New York, 1947.
The Discarded Image, CUP, 1961.

Mellor, J. W.: *Mellor's Modern Inorganic Chemistry*, revised and ed., G. D Parkes; Longmans, Green, London, 1945.

Newton, Sir Isaac: *Opticks*, Edition of 1730, Dover, New York, 1952.
Principia, Trans. I. Bernard Cohen and Anne Whitman, University of California Press, 1999.
Segrè, Gino: *Faust in Copenhagen*, Viking Penguin, 2007.
Smolin, Lee: *The Trouble with Physics*, Mariner Books, 2007.
Steiner, Rudolf: *Anthroposophy and Science*: Stuttgart 1921; Mercury Press, Spring Valley, NY, 1991.
Autobiography (The Story of My Life): SteinerBooks, Great Barrington, Mass. 1997.
The Boundaries of Natural Science: Dornach, 1920; Anthroposophic Press, Spring Valley, NY, 1982.
The Driving Force of Spiritual Powers in World History, Steiner Book Centre, North Vancouver, 1972.
Goethe the Scientist, Tr. Olin D. Wannamaker, Anthroposophical Press, New York, 1950.
Goethe's World View: Mercury Press, Spring Valley, New York, 1985.
Karmic Relations,Vol.1-8: Rudolf Steiner Press, London.
Knowledge of the Higher Worlds and Its Attainment, Anthroposophic Press, Spring Valley, New York, 1947.
The Light Course: tr. Raoul Cansino; Anthroposophic Press, Great Barrington MA, 2001.
The Origins of Natural Science: Stuttgart, 1921; Anthroposophic Press, Spring Valley, NY, 1985.
An Outline of Esoteric Science (1910): Tr. Creeger; SteinerBooks, Great Barrington, Mass. 1997.
The Philosophy of Freedom, tr. Michael Wilson; Rudolf Steiner Press, London, 1964.
The Riddles of Philosophy: Anthroposophic Press, Spring Valley, New York, 1973.

The Work of the Angels in Man's Astral Body, Zurich, 1918.

The Work of Secret Societies in the World - The Atom as Coagulated Electricity; Berlin, Dec. 23, 1904.

Stockmeyer, E. A. Karl: *Rudolf Steiner's Curriculum for Waldorf Schools*: Steiner Schools Fellowship, 1969.

J. R. R. Tolkien: *The Fellowship of the Ring*, Houghton Mifflin, Boston.

The Hobbit, Houghton Mifflin, Boston.

Tovey, Sir Donald: *Essays in Musical Analysis*, OUP, 1935.

Tyndall, John: *New Fragments*, D. Appleton and Company, New York, 1896.

About the Author

Keith Francis has a master's degree from the University of Cambridge, England. He worked as an engineer in the guided weapons department at Bristol Aircraft before entering the teaching profession and settling in New York City. He joined the faculty of the Rudolf Steiner School in Manhattan in 1965, and stayed there for over thirty years, teaching a variety of subjects, including physics, chemistry, mathematics, and music, and serving at different times as High School Administrator and Faculty Chair. His publications include *Death at the Nave*, and *The Education of a Waldorf Teacher*. His reviews and essays have appeared regularly in *Being Human* and *Southern Cross Review*. He is married, with two children, and divides his time between New York City and the Southern Berkshires of Massachusetts. Keith is in his fiftieth year as a member of the Anthroposophical Society.